THE EPITOME OF
ANDREAS
VESALIUS

THE EPITOME OF ANDREAS VESALIUS

TRANSLATED
FROM THE LATIN WITH
PREFACE AND INTRODUCTION BY
L. R. LIND

WITH ANATOMICAL NOTES BY
C. W. ASLING

FOREWORD BY
LOGAN CLENDENING

THE M.I.T. PRESS
MASSACHUSETTS INSTITUTE OF TECHNOLOGY
CAMBRIDGE, MASSACHUSETTS, AND LONDON, ENGLAND

TRANSLATOR'S DEDICATION

THIS BOOK IS DEDICATED
TO THE MEMORY OF
LOGAN CLENDENING
LATE PROFESSOR OF
THE HISTORY OF MEDICINE
UNIVERSITY OF KANSAS

TABLE OF CONTENTS

FOREWORD

WHEN Dr. L. R. Lind, Dr. H. C. Tracy, and Dr. C. W. Asling first came to us in the Department of the History of Medicine at the University of Kansas with the proposal to translate Vesalius into English, we were happy to be able to put our material at their service.

After careful discussion, it was generally agreed among us that the most practical plan would be to select the *Epitome* for complete translation. The *Fabrica* * by its bulk alone condemns itself, except for the hardiest and most curious scholars. What is true and sound in it has long since been incorporated in common anatomical knowledge.

Most of the mistakes have been discussed and are familiar to all who have undertaken more than a superficial review of Vesalius' place in the progress of science.

What would be useful, we all agreed, would be to place in the hands of the modern student a brief summary of what Vesalius said he was trying to do, what he found, and what he meant. Vesalius himself did this for us perfectly in the *Epitome*. Here, as Professor Lind says in the introduction, is a masterpiece of condensation. Here are the very words of the master himself, freed from controversial arguments long since outmoded. Here, indeed, is the *Epitome* of one of the greatest works on science ever to be written, one of the foundations of modern civilization.

The Department of Medical History is very proud to have had the privilege of sharing in the production of this work.

LOGAN CLENDENING
Professor of the History of Medicine

University of Kansas
November, 1944

* The *Fabrica* is now being translated by Professor J. B. deC. M. Saunders and Charles D. O'Malley.

ix

TRANSLATOR'S PREFACE

THE year 1943 marked the four-hundredth anniversary of the publication at the press of Johannes Oporinus in Basel, Switzerland, of both the *De Humani Corporis Fabrica Libri Septem* and the *Epitome* of that work. The *Epitome,* intended by Vesalius as a very brief descriptive anatomy and actually a remarkable condensation of the larger book, is here presented for the first time in a complete English translation; it has thus far been translated into no other modern tongues except German and Dutch.

The translation was begun at the suggestion of Professor H. C. Tracy, former head of the Department of Anatomy at the University of Kansas, who has served as a genial godfather to the undertaking. Doctor C. W. Asling, formerly of the Department of Anatomy, University of Kansas, has contributed the anatomical notes. I am greatly indebted to him for his patient and inspiring collaboration. It is a pleasure to acknowledge in this place the importance of his influence upon the entire book. His acute scrutiny has removed a number of errors from the translation; his concise, scholarly notes will make the text far more valuable to the medical student and historian than it might otherwise have been.

The late Logan Clendening made this translation possible by lending me his personal copy of the *Epitome* (1543); he showed a warm and generous interest in the progress of the work and was kind enough to write a Foreword for it. I deeply regret that he did not live to see the book in print. It is a great pleasure to express my appreciation to Mrs. Logan Clendening, who in pursuance of her husband's interest in the history of medicine made the publication of this book possible.

Doctor W. W. Francis, librarian of the Osler Library, McGill University, Montreal, Canada, has made several judicious suggestions from which I have profited and for which I thank him. I am indebted likewise to the kindness of Dr. Max H. Fisch, professor of philosophy at the University of

Illinois, Urbana, Illinois, for reading the manuscript and making a number of corrections.

The book has been read also by Professor J. B. deC. M. Saunders, chairman of the Divisions of Anatomy and of Medical History and Bibliography, University of California, San Francisco, California, who has helped me to avoid certain inaccuracies; for his assistance, so freely given, I am exceedingly grateful. His deep knowledge of the intricate history of sixteenth century anatomical terminology has been invaluable in setting me aright in many instances. Both he and Dr. Asling have persuaded me to use a more modern and technical phraseology throughout the translation, one more in keeping with the language now in use by the medical profession, rather than the somewhat picturesque English which Vesalius' Latin had at first tempted me to use. The difficulties involved in translating Vesalius are clearly represented by these words of Dr. Saunders in one of his letters to me. It should be remembered that they proceed from a scholar whose translations of the *Fabrica* and the *Venesection* and *China Root Epistles* of Vesalius are now eagerly awaited:

> I have never found Vesalius's style easy to handle but then neither did his contemporaries. Fabricius regarded it as difficult and Amatus Lusitanus calls it, with some justice, harsh. I seldom find a grammatical error but his Ciceronian periodic style is at times a trial and can be on occasions very obscure. This is particularly the case as he is attempting to use a technical language which is in the state of flux. I put in a long apprenticeship in developing a vocabulary of 16th century terms in medicine and have found the comparison of the Latin with the contemporary English translation of the works of Ambrose Paré most useful.

The translation has been completely rewritten in order to avoid the literalness to which both Dr. Francis and Dr. Fisch had also objected. I wish to thank Professor Henry E. Sigerist for his comments on the first sentence of Chapter I and for several encouraging letters.

Through the generous co-operation of Mr. A. Colish, and especially of Mr. Henry Schuman, to whom I am obliged for his unfailing concern and helpfulness in bringing the book to completion, I have been supplied with a photostatic copy of the contemporary (1543) German translation by Albanus Torinus, with which I have collated my translation throughout. Although I found Torinus sometimes in what I take to be error, his work cleared up several points for me—a fair exchange across four hundred years.

Dr. John F. Fulton, of the Historical Library, Yale Medical Library, has done me a signal honor in sponsoring this book as one of the Monograph Series published by the Historical Library. His constant and careful interest in the final stages of publication lays me under a debt difficult to repay.

I wish to thank both Mrs. Alta H. Lonnecker, secretary to the director of libraries, and Mrs. David D. Robb, formerly reference librarian, University of Kansas, for obtaining photostats, microfilm, and interlibrary loan books used in making the translation. To my wife, as often before, I am grateful for her capable assistance with the typing of the manuscript and the reading of the proofs.

L. R. LIND

Lawrence, Kansas
September, 1948

TRANSLATOR'S INTRODUCTION

THE LIFE AND WORK OF ANDREAS VESALIUS

I

THE echoes of the absurd quarrel between the sciences and the humanities in education still reverberate in certain quarters; but that quarrel has never had the slightest justification when the respective claims of each are considered from a historical point of view. Although Thomas Henry Huxley and Charles W. Eliot, each in his own land, fought hard to bring the sciences into the curriculum on a more equal footing with Greek and Latin, neither man wished to exclude the classical humanities in turn. They knew that in origin both forms of knowledge were embraced in the conception of the seven liberal arts; and they must have known that, in view of the basic unity of all knowledge, to speak of a "College of Liberal Arts and Sciences" was a tautological blunder.

Our age of specialization, with its urgent emphasis upon what is immediate, practical, and expedient, has been tempted in a fury of extremism to discard the classical humanities. Furthermore, while they are on the whole far more friendly to the Classics than their colleagues in certain other fields, some scientists, content to regard a moderate knowledge of French and German as exclusively sufficient for the linguistic training of their students, ignore the fact that the terminology of their sciences is predominantly Greek or Latin or both in its content. But those very scientists would be the first to recognize that an adequate understanding of the history of science is impossible without either a knowledge of Latin and Greek or the use of translations of the great books of any science written before the eighteenth century.

The best scientists of the Renaissance since Leonardo present an instructive contrast in this respect as well as in others to their modern brethren.

The former were not content to place limitations upon their own opportunities for wider knowledge with the simple excuse that there was not time for everything in a "crowded curriculum"; nor did they believe that one or two special talents, interests, studies, or languages were enough for their purposes. The scientist of that day was often an artist, a philologist, a philosopher, and a man of letters as well as a scientist. In fact, in Renaissance science, so little known to us as yet since many of its chief works, written in Latin, remain untranslated, the genuine rebirth of the ideals of classical civilization may be perceived sometimes even more clearly than in the more purely literary aspect of Renaissance culture. The Renaissance implies not only the rediscovery of Cicero but the revival of the Greek spirit of scientific inquiry in Europe. That rebirth of learning is typified in the persons of Leonardo da Vinci, Copernicus, Vesalius, and Galileo no less than in those of Petrarch and Erasmus; it has required the often neglected but patient and persistent efforts of the historian of science in our own day to bring this truth home to the modern humanist. These men of the Renaissance were profound enough to profit from, wise enough not to reject, and possessed of enough genius to build upon, the science, art, and literature of the ancient Greeks and Romans. Theirs was that marvelous unity of art, learning, and literature which has been lost in the narrow specialization of modern science; the bond which linked their common endeavors was a common understanding and respect for the tradition of ancient knowledge.

The year 1943 was the quadricentennial anniversary of two very important events in the history of science: the publication by Nikolaus Copernicus of his book *On the Motions of the Heavenly Bodies* and the publication by Andreas Vesalius of his work *On the Fabric of the Human Body.* That these two works, so far apart in subject, so close together in method and purpose, should have appeared simultaneously is not, perhaps, entirely a coincidence. Copernicus laid the foundations of modern astronomy and arrived at the true conception of the universe or the macrocosm; Vesalius, in his turn, founded modern anatomy upon a true conception of the physical microcosm, which is man. These ideas—the macrocosm and the microcosm—are among the most widely discussed in Renaissance thought; and it is no small part of the greater achievements of these scientists that they

brought these ideas at last out of the vague shadow of Platonic and Aris'
totelian dogmatism into the clear light of scientific reality. The dignity of
man, that favorite theme of Renaissance philosophers, gained further stature
from the detailed scientific discoveries of the age.

Andreas Vesalius of Brussels—anatomist, surgeon, teacher, philologist,
artist, philosopher, and poet—was a man worthy to be known as "un uomo
universale," in the soaring Renaissance meaning of the phrase. His achieve'
ments have been heralded from his day to ours in the highest terms, although
his books have been more often praised than read. He is the "founder of
human anatomy"; he "has left his name on the whole fabric of the human
body"; he is "the first modern anatomist to place his study on a firm founda'
tion of observation"; his book (*De Humani Corporis Fabrica*) is "not only
the foundation of modern Medicine as a science but the first positive achieve'
ment of Science itself in modern times." Sir William Osler, indeed, regarded
the *Fabrica* as the greatest medical book ever written.

Vesalius was born of a long line of medical men at Brussels on December
31, 1514. His family came originally from Wesel in Cleves and preserved
on its coat of arms the memory of that town by way of a pun in the shape
of three weasels; these appear also in the title pages of the *Fabrica* and the
Epitome. Vesalius' father was the personal apothecary of Charles V and fol'
lowed that monarch on the latter's campaigns until his own death in 1546.
After a youthful education at Brussels during which, as a sixteenth century
biographer writes, he dissected mice, moles, dormice, dogs, and cats, Vesalius
studied first at the University of Louvain and then at Paris from 1533 to the
autumn of 1536 under Jacobus Sylvius and Johannes Guinterius, both strict
Galenists in doctrine and method. Leaving a reputation for brilliance behind
him in Paris, he returned to Louvain in 1536. The next year found him again
in Brussels, but he soon left for Italy.

His departure from the North can probably be accounted for by the rela'
tively greater liberality of the intellectual atmosphere in the South. Certainly
in Padua he found for his pioneering efforts the sort of encouragement which
was soon to spur him on to the publication of his masterpiece, the *Fabrica*.
At Padua he was appointed to the chair of anatomy and surgery in the uni'
versity, a post he held until 1543. He passed his doctoral examination Decem'

ber 6, 1537, and assumed his teaching duties the next day when not yet twenty-three years old. At Padua he freed anatomical pedagogy from much of its antiquated apparatus by dissecting in person and dispensing with the ignorant barber-surgeon and the *ostensor,* the servant whose duty was to point out the parts of the cadaver while the lecturer, at a safe distance, explained them. In his courses in anatomy he cleared away many of the errors of Galenic anatomy. At Padua the greatest period in his productive life unfolded; here at the age of twenty-eight he completed the *Fabrica,* upon which he had been working for almost five years.

After visits to Venice and Ferrara, Vesalius journeyed to Basel, where the *Fabrica* and the *Epitome* were published by Johannes Oporinus, the famous humanist printer. At Basel, while visiting friends and seeing his books through the press, he found time to prepare a skeleton which he gave to the university; it can be seen there today. In September, 1544, he had married, at Brussels, Anna van Hamme, who bore him a daughter also named Anna. Vesalius took service as a court physician with Charles V from 1544 until he became physician (1556) to Philip II, to whom Charles had turned over the throne of Spain. Vesalius remained in Spain until 1564, practicing medicine at court and preparing a work on pathological anatomy which has not survived. His stay at the Spanish court is not well described for us in the sparse biographical details we have; he seems to have been despondent and to have lost interest in his science. In 1561 he wrote his last published book, a critique of Fallopius' *Observationes Anatomicae.*

After 1561 we begin to lose sight of Vesalius. His last writing is a diagnosis and prescription, one of a number which have been preserved, written at Christmas, 1562. In the spring of 1564 he appeared at Venice on his way to the Holy Land. Three separate traditions, all untrustworthy, profess to give the reasons for this pilgrimage; the only certainties are that he embarked upon it and on his return fell ill with plague or fever and died at Zante, in Greece, where he was buried by a Venetian goldsmith who had befriended him in his last days. Moritz Roth, in his meticulously documented biography, the best single book on Vesalius, briefly records the various tales, and their sources, concerning the end of his life. Charles Kingsley has summed up the matter in a few paragraphs of an essay on Vesalius:

After eight years of court life, he resolved early in the year 1564 to go on pilgrimage to Jerusalem.

The reasons for so strange a determination are wrapped in mystery and contradiction. The common story was that he had opened a corpse to ascertain the cause of death, and that, to the horror of the bystanders, the heart was still seen to beat; that his enemies accused him to the Inquisition, and that he was condemned to death, a sentence which was commuted to that of going on pilgrimage. But here, at the very outset, accounts differ. One says that the victim was a nobleman, name not given; another that it was a lady's maid, name not given. It is most improbable, if not impossible, that Vesalius, of all men, should have mistaken a living body for a dead one; while it is most probable, on the other hand, that his medical enemies would gladly raise such a calumny against him, when he was no longer in Spain to contradict it. Meanwhile Llorente, the historian of the Inquisition, makes no mention of Vesalius having been brought before its tribunal, while he does mention Vesalius' residence at Madrid. Another story is that he went abroad to escape the bad temper of his wife; another that he wanted to enrich himself. Another story—and that not an unlikely one—is that he was jealous of the rising reputation of his pupil Fallopius, then professor of anatomy at Venice. This distinguished surgeon, as I said before, had written a book, in which he had added to Vesalius' discoveries, and corrected certain errors of his. Vesalius had answered him hastily and angrily, quoting his anatomy from memory; for, as he himself complained, he could not in Spain obtain a subject for dissection; not even, he said, a single skull. He had sent his book to Venice to be published, and he had heard, seemingly, nothing of it. He may have felt that he was falling behind in the race of science, and that it was impossible for him to carry on his studies in Madrid; and so, angry with his own laziness and luxury, he may have felt the old sacred fire flash up in him, and have determined to go to Italy and become a student and a worker once more.

This is not the proper place in which to give more than a brief summary of the achievements of Vesalius, nor will medical readers need to be reminded of their importance. Roth's treatment of this subject is practically exhaustive. The writings of Vesalius, from the early revision of Rhazes to the examination of Fallopius, are characterized by a rigorous subjection of all medical and anatomical investigation to the evidence of observation, not to the authority of books or predecessors. The establishment of human anatomy upon a scientific basis is only one of his contributions to knowledge; he wrote also, as almost the first, upon pathological and comparative anatomy and on anthropology. Anatomical instruction was completely revolutionized by Vesalius; and his use of careful illustrations is a feature in it which modern teaching owes largely to him. The numerous discoveries he made in the "fabric" (or "working": the word has functional implications as well) of the human body are not distinguished as are those of other anatomists by the attachment of his name to any specific part of the body; but his revision of anatomy included more than two hundred corrections of Galen, as he states in the preface to the *Fabrica*.

The artistic activity of Vesalius is illustrated by the *Tabulae Anatomicae* of 1538, designed by himself, as were the illustrations of the *Fabrica* and *Epitome,* but executed by his countryman, Jan Stephan van Calcar. The illustrations of his books reveal a Renaissance feeling for symbolism carried out in the smallest details. The Classical influence appears in the cupids of the wood block—illustrated initials of the chapters, the children of the Asclepiad physicians of whom he speaks in his prefaces. The plates showing the external parts of the human body follow closely the classical statues of the Venus de' Medici and the Antinoüs. The frontispiece common to both *Fabrica* and *Epitome* represents a public anatomy conducted by Vesalius. Such public dissections were often gala affairs in the Renaissance, crowded with curious spectators and sometimes attended by the great. Certainly Vesalius seems to have been in his element performing before an admiring audience.

The philosophic bent of Vesalius was chiefly Platonic as is demonstrated by frequent references to Plato and by an unusual passage of Platonic doctrine in the *Fabrica* VII, 6, at the end of his description of the human brain. His pedagogical genius is apparent on every page of his books: the clear organization of material; the style of direct address to the student; the explicit, patient, and concise description—all indicate the gifts of a great teacher. The philological scholarship of Vesalius was exercised throughout his working life by his acute criticisms of ancient and contemporary anatomical doctrine, by his revision of Rhazes, and by his critical editorial work upon the Giunta edition of Galen, where the phrase in the heading of one of the pieces he revised—*aliquot in locis recognitus*—gives only a faint idea of the true magnitude of his contribution. His efforts in poetry may be gauged by the epigram "Ad Candidum Lectorem" in the *Epistola* of 1539 (Venesection Epistle), a competent but not inspired piece of humanist verse. His distich on the number of bones in the human body was plagiarized by several people and finally interpolated as verses 1627 and 1628 into the text of the famous medical poem, the *Schola Salernitana* (*Regimen Sanitatis Salernitanum*); see De Renzi, *Collectio Salernitana* V (Naples, 1859), 45:

> Adde quater denis bis centum senaque, habebis
> quam sis multiplici conditus osse, semel.

His motto, which appears beneath the 1542 portrait, was borrowed from the prose of Celsus: *Ocyus, iucunde et tuto.*

The personality of Vesalius himself remains, in spite of much self-revelation in his writings and biographical material of another nature, still largely an enigma. His far-ranging and penetrating intellect, his marvelous skill in controversy, his excellent Latin style, resemble those of Erasmus; indeed, Olschki calls him "the Erasmus of medicine." The customary and conventional flattery of his dedicatory prefaces does not conceal the real independence of Vesalius; he is one of the men whom the Renaissance princes delighted to honor for their genius and sought to attach to their courts: but he was never their vassal.

Anatomy, like astronomy, was a study always a bit suspected among churchmen. The procurement of cadavers was often a perilous business. Galen was the anatomical authority accepted by the Church. Yet Vesalius revolutionized anatomy, dissected cadavers without subterfuge or reticence in his public demonstrations, stole the bodies when necessary, and showed beyond a doubt that Galen knew anatomy only from the dissection of apes, dogs, and pigs—and did not thoroughly understand even the anatomy of these animals. The Church, nevertheless, did not lay hands upon Vesalius. Galileo and Servetus suffered under the Inquisition, but not Vesalius.

His portrait reveals an imperious, passionately energetic and self-willed character, confident, scornful, utterly fearless. The dark eyes, the short curly hair and beard, are physical features which recall other Renaissance portraits, but none perhaps is represented in so arresting an attitude as Vesalius, both in a large woodcut at the end of the text of the *Epitome* and in the frontispiece of the *Fabrica* and *Epitome*. Here, in probably the only authentic ones of the twelve different portraits which are preserved, Vesalius is shown as he demonstrates his art upon a cadaver identical with Figure 24 of the

Fabrica V. All the pride of his accomplishment, the self-assurance of his swiftly won knowledge, stand forth. The motley, eager crowd; the classic dissecting theater; the naked man whom he employed when pointing out the external parts of the body; lastly, the skeleton in the center with all its grim significance—these details are symbolic of his profession and reminders to his reader of the scenes amid which he achieved such signal success in the study of the human body. To this success, modern science owes the very foundation upon which it has erected the theory and practice of present-day anatomy.

THE EPITOME

II

THE *Epitome,* completed two weeks after the *Fabrica* on August 13, 1542, was published in June, 1543, and was followed two months later by the German translation of Albanus Torinus. The Latin *Epitome* which I have used consists of fourteen folio pages measuring twenty-one by sixteen inches. It contains eleven plates made from wood blocks, showing the bones, muscles, external parts, nerves, veins, and arteries; some portions of the plates were intended to be cut out and attached to their proper places upon the drawings of the body or skeleton. These plates show the human figure about 42.5 centimeters in height, thus revealing a proportion of one-fourth natural size if we reckon the average height of men as 165 to 175 centimeters. They were very probably executed by Jan Stephan van Calcar, possibly assisted, as in the pictures for the *Fabrica,* by some unknown artist and closely supervised throughout by Vesalius himself. Van Calcar completed the early *Tabulae Anatomicae Sex* of 1538, printed in Venice; but no artist is mentioned specifically by name by Vesalius in reference to the *Fabrica* and the *Epitome.*

In an age inspired in much of its art by the conception of death, when Holbein's great pictures of the dead and dying reflect the desolation and misery of plagues and massacres, it was only to be expected that Vesalius should present his anatomical figures in the gloomy attitudes of death. One skeleton figure leans like a Roman *genius mortis* upon a pedestal which bears upon it a pair of funereal verses from Silius Italicus (*Punica* XII, 243-44):

Solvitur omne decus leto, niveosque per artus
it Stygius color, et formae populatur honores.

The same plate in the *Fabrica* carries the sentiment, "Vivitur ingenio, caetera mortis erunt," instead. This figure "is said to have inspired Shakespeare's concept of Hamlet" (A. Castiglioni, *A History of Medicine,* 1941, p. 423). Although I have found no real evidence for this statement by Castiglioni, English literature plays a small part in the dissemination of Vesalian anatomy through Nicholas Udall, the author of "Ralph Roister Doister."

To the plagiarism in London of the Vesalian plates by Thomas Geminus (*Compendiosa Totius Anatomie Delineatio,* etc., English ed., 1553; 2d ed., 1559) an English text was provided, drawn not from Vesalius' Latin *Epitome* but from a text similar to that used by Thomas Vicary in his *Anatomie of the Body of Man* (1548). Udall seems to have done nothing more for this plagiarism by Geminus than to translate the *characterum indices* of the Vesalian plates and to re-arrange Vicary's text skillfully to conform with the anatomical teaching of Mundinus; he also added a preface. (See Sanford V. Larkey, "The Vesalian Compendium of Geminus and Nicholas Udall's Translation: their Relation to Vesalius, Caius, Vicary and de Mondeville": *Transactions of the Bibliographical Society,* XIII [London, 1933], 367-94; also, H. Cushing, *A Bio-Bibliography of Andreas Vesalius,* 1943, p. 127. Larkey also explains why Udall in the English Geminus abandoned the order of presentation of material in the *Epitome;* see below.)

The *Epitome* is, like the *Fabrica,* at once a descriptive anatomy and an anatomical atlas. Vesalius speaks of it most diffidently as an appendix, index, compendium, and pathway (*semita*) to the *Fabrica.* Clearly Vesalius wished to reach as wide a public as possible by means of the Latin and German texts of his manual, with the hope of forestalling as well as he could the plagiarists whom he knew were inevitable. The latter paid him their sincere compliments by pirating his books in England, France, and Germany.

The order in which the material is arranged in the *Epitome* differs from that of the *Fabrica*. This is apparent from the contents of each book as given below:

Epitome		*Fabrica*	
Book I	Bones and Cartilages	I	The Skeleton
II	Muscles and Ligaments	II	Muscles
III	Abdominal Viscera	III	Vascular System
IV	Thoracic Viscera and Vascular System	IV	Nervous System
V	The Brain and the Nervous System	V	Abdominal Viscera and , Organs of Reproduction
VI	Organs of Reproduction	VI	Thoracic Viscera
		VII	The Brain

The style of the *Epitome* is clear and brief. There is no evidence in it that Vesalius wrote bad Latin. Indeed, his style is among the best Latin styles written by the Renaissance thinkers; even the *Fabrica,* which involves an immensely greater amount of detail and is often couched in a conversational Latin natural between teacher and student, still preserves many classical features of Latin prose: the verb at the end of the sentence, the separation of subject and modifier by various forms of the verb, the careful arrangement of clauses, the use of the genitive, the clausula at the sentence-end. His choice of words is quite classical; yet, realizing the great necessity for systematizing the terminology of science (almost as bewildering in its present form as it was in his day), he constantly admits new words and phrases. The *Epistola* of 1539 shows a brilliance of epigrams, anecdotes, analogies, and descriptions which is not surpassed even by the books of his more mature years.

The *Epitome* is a triumph of condensation. Anyone who has examined the vast bulk of the *Fabrica* knows what an immense amount of detail has, in the *Epitome,* been reduced to the lowest possible limits. Written in language which does not merely repeat that of the *Fabrica,* the *Epitome* is a book in its own right, independent in treatment, point of view, and purpose. The book embodies the principles of his educational method in a more striking fashion than does the *Fabrica.* Not the least important of his teaching devices is shown in the plate which calls for the clipping out of certain anatomical features and their superimposition by pasting upon the larger figure of the human body.

The tone of the *Epitome* is sober and impartial; Vesalius never mentions his contemporaries by name and scarcely refers even to Galen, except in the dedicatory epistle. "Professors of dissection" is a term he uses when necessary; the comments in the context where those words appear are usually Vesalius' own views upon controversial points and, by a charac-teristic ironical implication, he manages to give the impression that the "pro-fessors" are wrong. Naturally the students for whom the *Epitome* was in-tended would not be interested in scholarly controversy; hence Vesalius does not indulge in it. His purpose was pedagogical; the *Epitome* was a guide, a brief manual, an index, as he called it, to the parts of the body; and, master that he was, he never departs for a moment from that purpose. Sel-dom has so large an amount of scientific knowledge been so skillfully com-pressed into the narrow limits of a few pages.

The monumental *Bio-Bibliography of Andreas Vesalius*, the result of more than forty years of devoted study by Harvey Cushing (New York: Schuman's, 1943), is the great mine of information to which all readers of this first complete published translation into English of any of the longer works of Vesalius must now be referred for additional facts about the edi-tions and plates of his books as well as for certain biographical material. The bibliography there provided is exhaustive, elaborately detailed, and sumptu-ously printed. In this brief introduction I have intended to convey merely the chief facts about Vesalius and the *Epitome*. The very short bibliography immediately following contains a selection of the more important works on Vesalius and his writings from which the student of Renaissance medical history and the general reader would be most likely to profit.

The reader's attention is called to the list of marginal notes by Vesalius given at the end of the text of this translation. This is a series of Greek equivalents for the Latin terms used by Vesalius in his text; he did not, as will be seen, make it either exhaustive or carry it out through the six books with the same consistency. These Greek terms, occurring as they do with great frequency in modern anatomical terminology, will help the reader to find his way more easily and provide additional synonyms for the Latin and English terms.

Dr. Asling's general interpretations for each book together furnish a convenient orientation for understanding the anatomical principles discussed in the *Epitome*. They sum up briefly and competently the substance of each book and are particularly recommended to the reader of this translation. Both the general interpretations and serial notes often serve as a condensed paraphrase of the text such as was provided by Albanus Torinus in his German translation of the *Epitome*. (See Henry E. Sigerist, "Albanus Torinus and the German Edition of the *Epitome* of Vesalius"; *Bulletin of the History of Medicine*, XIV [1943], 652-66.) Albanus, faced by the lack of equivalent German scientific terms, was forced to coin words and to resort to paraphrase. Modern English possesses most of the required terms for reproducing the Latin of Vesalius. I have not often therefore resorted to paraphrase in translating the *Epitome* but have attempted to say as clearly and briefly as possible in English what Vesalius said so well in Latin.

L. R. LIND

BIBLIOGRAPHY

THE PRINCIPAL EDITIONS OF VESALIUS

Paraphrasis in nonum librum Rhazae medici Arabis clariss. ad regem Almansorem de singularum corporis partium affectuum curatione, autore Andrea Wesalio Bruxellensi medicinae candidato. Lovanii ex officina Rutgeri Rescii. Mense Februar. 1537.

Ibid., Basileae in officina Roberti Winter. Anno 1537. Mense Martio.

[Tabulae anatomicae.] Imprimebat Venetiis B. Vitalis Venetus sumptibus Joannis Stephani Calcarensis. Prostrant vero in officina D. Bernardi. Anno 1538.

Institutionum anatomicarum secundum Galeni sententiam ad candidatos medicinae Libri quatuor, per Joannem Guinterium Andernacum medicum. Ab Andrea Wesalio Bruxellensi, auctiores et emendatiores redditi. Venetiis in officina D. Bernardini. 1538.

Andreae Wesalii Bruxellensis, scholae medicorum Patavinae professoris publici, Epistola, docens venam axillarem dextri cubiti in dolore laterali secandam: et melancholicum succum ex venae portae ramis ad sedem pertinentibus, purgari. Basileae, in officina Roberti Winter. Mense Aprili. Anno 1539.

Galeni omnia opera nunc primum in unum corpus redacta. Apud haeredes Lucæantonij Juntae Florentini Venetiis. 1541. Vol. II:

Galeni de nervorum dissectione liber ab Antonio Fortolo Joseriensi latinitate donatus, et ab Andrea Wesalio Bruxellensi aliquot in locis recognitus.

Galeni de venarum arteriarumque dissectione liber ab A. F. Joseriensi latinitate donatus, et ab Andrea Wesalio Bruxellensi plerisque in locis recognitus.

Galeni de anatomicis administrationibus libri novem ab Joanne Andernaco latinitate donati, et nuper ab Andrea Wesalio Bruxellensi correcti, ac pene alii facti.

Andreae Vesalii Bruxellensis, scholae medicorum Patavinae professoris, de humani corporis fabrica Libri septem. Basileae, ex officina Joannis Oporini. Anno salutis reparatae 1543. Mense Junio.

Andreae Vesalii Bruxellensis, invictissimi Caroli V. Imperatoris medici, de Humani corporis fabrica Libri septem. Basileae, ex officina Joannis Oporini, Anno salutis per Christum partae 1555. Mense Augusto.

Andreae Vesalii Bruxellensis, scholae medicorum, Patavinae professoris, suorum de humani corporis fabrica librorum Epitome. Basileae, ex officina Joannis Oporini. Anno 1543. Mense Junio.

Von des menschen cörpers Anatomey, ein kurtzer, aber vast nützer ausszug, auss D. Andree Vesalii von Brussel Bücheren, von ihm selbs in Latein beschriben, unnd durch D. Albanum Torinum verdolmetscht. Gedruckt zü Basel, bey Johann Herpst, genant Oporino, unnd vollendet am neünten tag des Augstmonat, nach der geburt Christi imm 1543 Jar.

Andreae Vesalii Bruxellensis, medici Caesarei epistola, rationem modumque propinandi radicis Chynae decocti, quo nuper invictissimus Carolus V. Imperator usus est, pertractans: et praeter alia quaedam, epistolae cuius' dam ad Jacobum Sylvium sententiam recensens, veritatis ac potissimum humanae fabricae studiosis perutilem: quum qui hactenus in illa nimium Galeno creditum sit, facile commonstret. Basileae, ex officina Joannis Oporini, anno salutis humanae 1546. Mense octobri.

Andreae Vesalii, anatomicarum Gabrielis Falloppii observationum Examen. Venetiis, apud Franciscum de Franciscis, Senensem. 1564.

Collected Works:

Andreae Vesalii Invictissimi Caroli V. Imperatoris Medici Opera omnia Anatomica et Chirurgica cura Hermanni Boerhaave. . . . et Bernhardi Siegfried Albini. Lugduni Batavorum, apud Joannem du Vivie, et Joan. et Herm. Verbeek, 1725. 2 vols.

Andreae Vesalii Bruxellensis Icones anatomicae. Ediderunt Academia

Medicinae Nova-Eboracensis et Bibliotheca Universitatis Monacensis, 1934. (Series of plates from original woodcuts of *Fabrica* and *Epitome*.)

BOOKS AND ARTICLES ABOUT VESALIUS

Ball, J. M., Andreas Vesalius, the Reformer of Anatomy. St. Louis: Medical Science Press, 1910.

Boyden, E. A., "The Problem of the Double Ductus Choledochus," etc. [a translation of gall-bladder description from the *Fabrica*], *Anat. Rec.,* LV (1932), 74.

Burggraeve, A., Etudes sur André Vésale, précédées d'une notice historique sur sa vie et ses écrits. Gand, Belgium: C. Annoot-Braeckman, 1841.

Castiglioni, A., A History of Medicine. New York: Alfred A. Knopf, Inc., 1941, pp. 76, 418-27, 431, 444, 490.

Choulant, L., History and Bibliography of Anatomic Illustration; tr. with preface by M. Frank. Chicago: University of Chicago Press, 1920.

Clendening, Logan, Source Book of Medical History. New York: Paul B. Hoeber, Inc., 1941, pp. 126-51. Contains translation of the preface to the *Fabrica;* also a translation of "On Dissection of the Living," *Fabrica,* Book VII, xix.

Cushing, Harvey, A Bio-Bibliography of Andreas Vesalius. New York: Henry Schuman, 1943.

Feyfer, F. M. G. de, "Die Schriften des Andreas Vesalius." *Janus,* XIX (1914), 435-507.

Fisch, Max H., "Vesalius in English State Papers." *Bull. Med. Libr. Ass.,* XXXIII (1945), 231-53.

Foster, Sir M., Lectures on the History of Physiology. Cambridge: The University Press, 1901.

Kingsley, Charles, "Vesalius, the Anatomist." In his: Health and Education. New York: D. Appleton and Company, 1893, pp. 385-411.

Lambert, S. W., and Goodwin, G. M. Medical Leaders from Hippocrates to Osler. Indianapolis: Bobbs Merrill Co., 1929.

Morley, Henry, "Anatomy in Long Clothes." *Frasers Magazine,* Nov., 1853.

Olschki, Leonardo, Geschichte der neusprachlichen wissenschaftlichen Literatur I (Leipzig, 1919), 265, 284; II (Leipzig, 1922), 16, 24, 34 ff., 39, 81 ff., 95, 98 ff., 178, 330.

Roth, Moritz, Andreas Vesalius Bruxellensis. Berlin: Georg Reimer, 1892.

Saunders, J. B. deC. M., and O'Malley, C. D., The Bloodletting Letter of 1539. In: Studies and Essays in the History of Science and Learning offered to George Sarton, ed. by M. F. Ashley Montagu. New York: Henry Schuman, 1946, pp. 5-74.

————, "A Reading from the De Humani Corporis Fabrica of Andreas Vesalius." *J. Amer. Coll. Dentists,* X (1944), pp. 211-18.

————, "Bernardino Montana de Monserrate, Author of the First Anatomy in the Spanish Language; Its Relationship to De Mondeville, Vicary, Vesalius, the English Geminus, and the History of the Circulation." *J. Hist. Med. Allied Sci.,* I (1946), 87-107.

Singer, Charles, The Evolution of Anatomy. London: Kegan Paul, Trench, Trubner & Co., 1925, pp. 111-35.

Spielmann, M. H., The Iconography of Andreas Vesalius. London: John Bale, Sons and Danielsson, Ltd., 1925.

VESALIUS TO THE READER *

THE compendium of the books on the fabric of the human body which I now publish is divided into two sections, of which the first is comprised of six chapters embracing a most succinct description of all the parts; the second section displays their delineation in a number of plates accompanied by indices of the characters by which they are designated. Therefore, it is a matter of your own choice and dependent upon your desire whether you shall first approach my ordering of the material (which from various considerations of printing and drawing I have selected as the most suitable) from the description of the parts, or their designation and index of their distinctive features starting with the figures showing the nude forms of man and woman, where the names of their external portions, serving as an index to these figures, occur.

The figure printed on the recto of the leaf with the nude male figure offers the complete structure of the bones,† whereas the figures drawn to the same scale as the nude man and labeled as muscle tables show the bones also in the order in which they are reached in dissection, especially the fourth and fifth muscle figures.‡

The delineation of the muscles and ligaments is to be sought first from that figure which we place opposite the one showing all the bones, and it is therefore called the first muscle figure. Next in order comes that which is called the third, then the fourth and fifth.§ The organs by which

* From the frontispiece of the *Epitome*. An English translation appears also in H. Cushing, *A Bio-Bibliography of Andreas Vesalius*, New York: Schuman's (1943), pp. 110-11, where an attempt has been made to indicate by symbols the exact references to the plates. Vesalius is not entirely clear in his suggestions for the use of the plates; an error, moreover, exists in the serial enumeration, thus making the task of comparing the text with the plates more difficult. On this entire matter see W. G. Spencer, "The 'Epitome' of Vesalius on vellum in the British Museum Library"; *Essays on the History of Medicine presented to Karl Sudhoff*, etc., edited by Charles Singer and Henry E. Sigerist, Zürich, Seldwyla (1924), pp. 237-44.

† That is, the skeletal figure, while it shows all the bones, is drawn to a smaller scale since it was taken from the *Fabrica*, where the plates are smaller.

‡ The plates are not numbered; if they were, these would be two and one respectively, not four and five.

§ For the reason why Vesalius here skips the second muscle figure, see W. G. Spencer, *op. cit.*, and especially Larkey, *op. cit.*

nutrition with food and drink is maintained, then the heart and the parts subserving its functions, together with the nervous system, are shown in figures on the plates following that of the nude woman. Therein are also to be seen the female organs of generation, and likewise the male organs occur in a picture to be cut out and pasted to the fifth muscle figure.

As for the portrayal of those parts which are contained in the skull, what is not carried by the plate prepared for demonstrating the nerves is presented in sufficient fullness in the plates showing the muscles. These should be taken in the following order: the head of the first figure, then of the second, then of the fourth, together with the images which the hands of that figure hold; next, that image which appears in the left hand of the fifth figure and the other image lying on the ground beyond the picture of the parts of the eye. Farewell, and use these, my efforts, as frankly as they are offered.

THE EPITOME OF HIS BOOKS ON THE FABRIC OF THE HUMAN BODY BY ANDREAS VESALIUS OF BRUSSELS

TO THE MOST SERENE PRINCE PHILIP, SON AND HEIR
OF THE DIVINE EMPEROR CHARLES THE FIFTH,
MIGHTIEST AND MOST INVINCIBLE
ANDREAS VESALIUS SENDS GREETINGS:

WITHIN the slender compass of these pages, greatest prince Philip, adorned with the splendor of your immortal name and under its good auspices, there is sent forth unto the common use of learning the description of the human body, which I have so divided in the manner of an enumeration, and singly related, that the principal branch of natural philosophy, treating the finished product of a creation most perfected and nigh the most worthy of all, may in the manner of an image be set before the eyes of those studious of the works of Nature. This has been done with as much conciseness as possible, and with less labor it describes those matters which I have embraced more amply in my seven books upon the subject. To those books this *Epitome* is, as it were, a footpath, or, as it will also be rightly considered, an appendix, gathering into summary form the chapters which are set forth with detail in those books; it lays out everything in such a fashion I may prophesy that you, with the amazing liberality of culture in which you eagerly welcome whatsoever slightest offering of the writer's craft, will not utterly cast it from your sight. Moreover, as you are now entering upon a period in your life distinguished by such various virtues, you are held fast by a wondrous and most generous love of all art and learning. And when your spacious spirit shall one day rule the whole world, you may perhaps at times consider it pleasant to be acquainted with my work and to regard it as a situation wretched and unworthy of the greatest Emperors, Kings, and Consuls, that in the pursuit of studies so varied, the harmony of the human body which we shall publish to the world should lie constantly concealed; that man be completely unknown to himself; and that the structure of instruments so divinely created by the Great Artificer of all things should remain unexamined: since it is by the function of these instruments that those things we look upon as most, and almost solely, important are brought to pass.

Truly, although for this reason my undertaking will perhaps be not wholly displeasing to your admirable judgment, if I should nevertheless refuse to give forth this companion to physicians because, while I strive to be useful to them yet at the same time I am anxious to snatch opportunity from the hands of certain rascally printers who may later seize in possession upon the labors of another to reduce them ineptly into small space and publish them under their own names (creatures born for the destruction of letters!), I might in either case prove a grievous hindrance. For no one is ignorant how much is lost in all sciences by the use of compendiums. Though indeed they seem to provide a certain way and systematic approach to the perfect and complete knowledge of things and seem to contain in short and in sum that which is set down elsewhere with more space and prolixity and are for this reason considered in the light of an index or the very abode of memory, in which matters written down at length are fitly reduced to their proper place, nevertheless, compendiums do signal injury and wreak a great havoc upon literature; for, given to the use of compendiums alone, we read scarcely anything else through to the end these days. This is true even for those who have delivered themselves completely to learning, to this degree aspiring only to the shadow and superstructure of science, digging little or not at all beneath the surface.

However, although this evil wanders widely amid almost all studies, it is a charge to be laid most gravely at the door of the mob of physicians that they perform their duty so carelessly in distinguishing the parts of the human body that not even enumeration is made use of in learning them. For when, beyond the function and use of each part, its location, form, size, color, the nature of its substance, the principle of its connection with the other parts, and many things of this sort in the medical examination of the parts may never be sufficiently perceived, how many can be found who know even the number of the bones, cartilages, ligaments, muscles, and veins, arteries, and nerves running in a numerous succession throughout the entire body and of the viscera which are found in the cavities of the body? I pass over in silence those pestilent doctors who encompass the destruction of the common life of mankind, who never even stood by at a dissection: whereas in the knowledge of the body no one could produce anything of value who did not perform dissections with his own hands as the kings of Egypt were wont to do and in like manner busied himself frequently and sedulously with dissections and with simple medicines. Whence also those most prudent members of the household of Asclepius will never be sufficiently praised, who, as children in the home learn reading and writing, so they exercised the dissection of cadavers and, learned in this wise, under the happy auspices of the Muses, they bent to their studies. Furthermore, whatever our sloth in the thorough mastery of Anatomy as the basis and foundation of the medical art, I have assumed that no demonstration is required of how necessary the knowledge of human parts is for us who have enlisted under the banner of medicine, since the conscience of each and all will bear full testimony to the fact that in the cure of illness the knowledge of those parts lays rightful claim to first, second, and third place; and this knowledge is to be sought primarily from the affected portion, without, of course, neglecting the due application of subsidiary remedies. Indeed, those who are now dedicated to the ancient study of medicine, almost restored to its pristine splendor in many schools, are beginning to learn to their satisfaction how little and how feebly men have labored in the field of Anatomy to this day from the times of Galen, who, although easily chief of the masters, nevertheless did not dissect the human body; and the fact is now evident that he described (not to say imposed upon us) the fabric of the ape's body, although the latter differs from the former in many respects.

But as to my own audacity, by virtue of which this slight offering, unworthy of your majesty and uniquely commended by such a patronage, hazards the dubious fortune of critical judgment, I shall defend myself with no excuse except that this is the grain and salt whereby I am permitted to obtain favorable omens for my systematic studies; and meanwhile I should wish this book to be an indication of my complete obedience and sense of duty toward my country's ruler until such time as it shall be possible to offer incense also.

At Padua, on the Ides of August, in the Year of the Virgin Birth,
MDXLII. [August 13, 1542]

THE EPITOME OF

ANDREAS
VESALIUS

CONCERNING THE BONES AND CARTILAGES,

OR THOSE PARTS WHICH SUPPORT THE BODY

CHAPTER I

ALL of the parts of the human body are either *similar,* or simple to the senses,* such as the bones, cartilages, ligaments, fibers, membranes, flesh, and fat, or *dissimilar,* instrumental agencies, such as the veins, arteries, nerves, muscles, individual fingers, and the remaining organs of the entire body. These latter parts are made all the more instrumental to the degree that they (e.g., the hand and head) are composed of many similar parts and also of many functional mechanisms [1].

The bones are the hardest and driest parts of the entire body. The cartilages are softer than the bones, but of the parts of the body they are next in hardness to the bones. The whole body is supported by bones and carti-

---

* I am indebted to the kindness of Professor Henry E. Sigerist, former director of the Institute of the History of Medicine, Johns Hopkins University, for the following answer to my question about this passage: "I think that *sensu ue* refers to the senses of the anatomist, that is to say, these parts are simple as judged by the senses; the anatomist cannot perceive any simpler parts composing them. For this interpretation we have a good parallel in Galen, ed. Kühn, Vol. 4, p. 741: εἴη ἂν οὖν ἀρίστη κατασκευὴ τοῦ σώματος, ἐν ᾗ ᾗ τὰ ὁμοιομερῆ πάντα (καλεῖται δ' οὕτως δηλονότι τὰ πρὸς αἴσθησιν ἁπλᾶ) τὴν οἰκείαν ἔχει κρᾶσιν. Vesalius' 'sensu ue simplices' stands for πρὸς αἴσθησιν ἁπλᾶ, which expression the Latin translator of the Galenic text renders by: 'sensu simplices.' I should therefore be inclined to translate the beginning of the sentence: 'All parts of the human body are either similar or simple to the senses as bone, cartilage. . . .'"

Professor Sigerist quotes both the Latin text and the translation of this passage into German by Albanus Torinus in his article "Albanus Torinus and the 'Epitome' of Vesalius," *Bulletin of the History of Medicine* XIV (1943), 663-64, and comments on the virtues of each passage.

Professor Saunders writes me as follows: "The classification of the parts of the body into 'similars and dissimilars' was standard terminology almost to the time of Bichat. The terms are of course derived from Galen. What is meant is parts of similar temperaments and elements. This passage is typical of one of the difficulties of the modern translator when dealing with the humoral doctrine; the very terms have disappeared from our language with the development of newer anatomical and physiological concepts even in everyday language. Contemporary sixteenth century English is as above: 'similars and dissimilars.' Cf. animal, spiritual, natural—all terms with special meaning differing from that of customary Greco-Latin usage. The same difficulty as above is presented with 'functional.' 'Instrumentum' is not quite 'functional.' The term is derived from Aristotle, *De Gen. Animal.* I, ii, 716a: ' . . . the bodily parts are the instruments or organs to serve the faculties. . . .' According to Galen, *Method, Medendi* I, vi, instruments are those parts of the body capable of carrying out a complete action; hence the arteries, veins, and nerves are collectively instruments for the distribution of the vital, natural, and animal spirits; the eye, and all pertaining to it, the instrument of sight. This was the customary usage of the *Fabrica.*"

1

lages in combination, and all its parts are attached to, and stabilized by, them. The skull, which contains the brain and the sense organs, is composed of several bones; one bone frequently makes up the forehead, especially in women [2]; in the occiput of the head there is likewise one; two bones are at the vertex, and one at each ear or temple [3], in which a cavity is formed suitable for the reception of the organ of hearing. This cavity encloses two small bones, one of which resembles an anvil or molar tooth, while the other is like a small hammer [4]. The bone of the temple has connected to it the cartilage supporting the structure of the ear. In addition it has three processes: first, it bears the mastoid process posteriorly; second, the styloid process, like a bodkin or cock's spur; third, a process which turns anteriorly, is closely joined in the external structure of the orbit to the bone of the upper jaw, and at the level of this junction constitutes the part of the skull which we call the os jugale [5]. Inferiorly the temporal bone joins the base of the occipital bone to form that region of the skull which we may compare to a broken rock, not only from the standpoint of hardness but also in appearance [6]. In the base of the skull is a noteworthy bone which professors of dissection compare to a wedge [7]; it is polymorphous and gives forth projections very much like the wings of a bat [8]. Near it, above the nose, there is another bone which is pervious in form like a sieve or more like a sponge [9]; it forms the septum of the nose, and in company with the seven bones previously listed it makes up the cavity which contains the brain.

In the upper jaw we count twelve bones in addition to the teeth: one on the lateral and two on the medial aspect of each eye socket [10], and one pair in the inferior part of the orbits; these last bones are regarded as by far the largest of the group [11] and contain small cavities for the reception of the upper teeth. A pair of bones is found at that end of the palate at which the apertures of the nasal cavities become continuous with the pharynx [12]. Finally, the prominent part of the nose is composed of two bones which connect with the cartilages of the nose by which the wings of the nostrils are supported; like the other bones of the upper jaw, they have no special name [13].

Thus far most of the bones mentioned are held in place by sutures, of which the one passing transversely in the sinciput [14] is called the coronal; the one creeping transversely in the occiput receives its name from its resemblance to the Greek letter Λ [15]; and the one passing along the length of the skull from the top of the occiput to the middle of the coronal suture is called the sagittal. Those bony junctions which are equidistant from the sagittal suture and which pass superior to the ears do not look like sutures but are more like two overlapping scales, from which appearance they are called squamous conglutinations [16]. The remaining connections between the bones of the skull do not show the appearance of a suture as well as do the three which have been named; in fact, in some places they appear more like a simple line and should receive the name of "harmony" rather than that of "suture" [17].

The lower jaw consists of a single bone, with the exception of very young children, in whom it forms from two bones fusing at the point of the chin; indeed, most of the bones of children are composed of several bones which later unite into single bones in those who have reached their full growth [18]. The lower jaw articulates on either side with a bone located near the ear, and in addition with its special cartilage intervening here, where the heads of these bones and their corresponding sinuses fit together [19], the articular surface is covered in the manner of a rind and a joint is thereby provided that turns readily, rendering it free from the injuries of continuous friction of the bones against each other. There are at most sixteen teeth in each jaw: four incisors, two canines, and ten molars; not only do they differ in appearance when examined in place in the mouth, but they are also fixed in their sockets with differing numbers of roots.

In the pharynx, the bone at the root of the tongue has more the shape of the Greek letter υ than of Λ [20]; it is bound together with many small bones. The lower of these are joined at their extremities to the rough artery of the head [21] (which we call the larynx rather than the gullet or throat) and, more specifically, to a cartilage which resembles a shield. This cartilage is called the first cartilage of the larynx and can be palpated in its entirety [22]. The second cartilage makes up the greater part of the posterior portion of the larynx and resembles the ring which the Thracians fit on their right thumbs when shooting arrows [23]; it lacks a name and from this fact

3

actually receives one [24]. The third cartilage, which is made up of two parts, resembles the orifice of vessels used for pouring water in washing the hands [25]; it forms an opening in the middle of the larynx for structures like the reeds of a flute [*tibia*] or pipe [*fistula*] [26]. This opening is like the aperture of pipes which are held to the mouth, and because of this we call it the "little reed" [27]. On these cartilages an operculum is laid like a lid, composed of cartilage [28], fat, and ligamentlike structures. The remaining cartilages of the rough artery, making up both its trunk and the branches running to the lungs, present the semicircular shape of the letter C or the Greek sigma [29].

The backbone, which provides the best passage for the dorsal medulla [30], is like a keel of the body and is divided into the regions of the neck or cervix, the thorax, the loins, the sacrum, and the coccyx or cuculus [31]. It is composed of thirty-four bones, which we call vertebrae. The neck has seven bones, furnished with several processes (not all bones having the same number); by means of the first of these bones [32] (which, unique among the vertebrae, has no spine and has quite protuberant transverse processes) we move the head directly forward and backward. By the use of the second vertebra (to which a prominent process resembling a canine tooth is attached) we turn the head [33]. By the function of the remaining vertebrae the head is carried to the side but only in a slight degree.

The thorax has twelve vertebrae, with which the ribs articulate. Usually the lowest of these vertebrae [34] is supported by the adjacent vertebrae through its ascending and descending transverse processes (with which the vertebrae are articulated in turn in addition to the connections between their bodies) just as the first cervical vertebra supports the bones contiguous to it above and below. The remaining vertebrae of the backbone above the twelfth thoracic are supported from above and in turn support those below, while those below this level are supported below and support those above them. Below the twelfth vertebra are five vertebrae of the loins, followed by the os sacrum, formed, at most, from its own six bones firmly fused together. The coccyx is a bone the extremity of which terminates in cartilage; it is frequently composed of four meager ossicles which provide no foramina for nerves and have no path for the dorsal medulla.

Furnishing a suitable location for the heart and the organs subsidiary to the heart, the breastbone occupies the anterior region of the thorax. It is composed very rarely of seven bones [35], usually fewer, receiving the seven articulations of the ribs on either side. Of the twelve ribs on each side ending in cartilage, the seven upper ones attach to the breastbone by their own cartilages and therefore have been given the name of true or genuine ribs. The ribs which do not come in contact with the breastbone, but fall short of it and of the anterior region of the abdomen in a degree corresponding to the lower position which they occupy, are called false ribs [36]. The lower portion of the breastbone terminates in cartilage and resembles the blunt point of a sword (because of which the entire breastbone is compared to a sword) [37]. At the upper extremity of the bone, where it is broadest and strongest [38] and where the throat lies, a clavicle is attached on each side, holding the shoulder joint away from the thorax.

The scapula has the form of a triangle and occupies the posterior region of the thorax on either side; it ends in a neck in which a socket is molded [39], suitable for the reception of the head of the humerus. From its own back the scapula sends out a process projecting above the shoulder joint and called the summus humerus [40]. It is articulated through a special cartilage of the clavicle (as is also the case with the connection between the clavicle and the breastbone); by means of the clavicle the shoulder joint is held away from the thorax. The inner process of the scapula is compared to an anchor, or the letter C, or the Greek sigma [41].

The bone of the arm, or humerus, articulates with the scapula; inferiorly it is furnished with several sinuses and tuberosities [42]. Two bones are joined here, the radius and the ulna (the other name of which is cubitus, from the name of the entire member). The cubitus is flexed and extended on the humerus in a fashion similar in both man and quadrupeds. In its upper portion, in the posterior aspect of the arm joint, it has a process which many call the gibber [43]. Similarly, its lower portion has another process which takes its name from the shape of a stylus.

The wrist articulates almost exclusively with the radius and is separated from the ulna by its own cartilage [44]; it is made up of eight bones which are entirely different from each other in shape and size. The four bones of the postbrachial metacarpals, together with the first bone of the thumb, join the wrist at its lower aspect. The thumb, as well as the other four fingers, is composed of three bones placed serially, as it were, in a line [45]. The total number of bones of the fingers is fifteen; to this we add the two small bones comparable to sesame seed [46] which are found at the second inter-node of the thumb. Bones which are similar, but much smaller and quite cartilaginous, are found in the inner portion of the first four internodes of the fingers and in the third internode of the thumb; one connected near the wrist to the external side of the metacarpal bone which supports the little finger is found by those who dissect the body. There is rarely one in the remaining internodes of the fingers, except that such may be observed in the fingers of very old men [47].

A large bone is attached on each side of the os sacrum; where this bone is broad and faces the flanks, it is called the ilium, but where the head of the femur enters a deeper socket, it is called the hipbone. That aspect of the bone which, with the bone of the opposite side, forms the pubis (which has a conspicuous perforation) is called the pubic bone. Henceforth, however, under the name of coxendix we include the entire bone [48].

The rounded head of the superior aspect of the femur is continuous with the broad neck; it grows inward in an oblique direction and articulates with the coxendix. At its inferior end, the femur enters the sinuses [hollows] of the tibia by means of its own two heads [condyles], with a certain sinus lying between them where a tuberosity of the afore-mentioned tibia is received, as well as by means of the special cartilages in this joint, which is very similar to the knee of quadrupeds and of birds [49]. A large process is seen near the external part of the neck of the femur, called the rump or the great pulley [rotator] [50]. Inwardly, the bone has also another process, much smaller than the one on the external aspect; because of this fact we call it the minor or internal rotator.

In the leg, just as in the forearm, two bones are seen; the inner of these is much thicker than the outer one and goes by the name of the entire member [51]. The outer bone, which is not articulated to the femur, is called the fibula. A round bone called the mola or patella, similar to a shield, is located in front of the joint of the tibia with the femur. The lowest portions of the tibia and fibula, which have prominences on their outer aspects protruding from the fleshy parts, are called the malleoli.

The talus lies in the same region in man and in beasts and is put to the same use in both creatures. Under it lies the os calcis [52], extending posteriorly far beyond the straightness of the tibia. The anterior part of the talus ends in a round head; it enters the sinus of the navicular bone, which is joined to three bones of the tarsus [53]. The fourth, an external bone of the tarsus, resembles a cube or a die and adjoins the heel. Five metatarsal bones [*ossa pedii*] [54] are joined with the four bones of the tarsus; by means of these five the toes are supported.

The great toe is formed from two internodes [55], while three are assigned to each of the remaining toes. Moreover, there are the same number of these bones in the foot (which is much shorter in man than in quadrupeds) and in the hand. Small bones occur, which we compare to sesame seed [56]; the two which lie beneath the first internode of the great toe are far larger than in the hand. Of these, the inner may be the one which followers of occult philosophy affirm to be subject to no sort of corruption; foolishly they contend that it must be long preserved in the earth until, in the manner of a seed, it produces a man at the time of resurrection [57]. To the bones of the digits, in the foot as well as in the hand, nails are attached; we refer to them here not inopportunely on account of their substance, just as among the parts bracing something we mention the cartilages of the eyelids, supporting the lids so that they may not fall together flaccidly.

MUCH has been written about the splendid title pages of Vesalius' *Fabrica* and *Epitome*. Some see them as representing symbolically the divisions of Vesalian anatomical teaching into three phases: the skeleton, in its central position of dominance, represents the foundation of most anatomy teaching, osteology; the cadaver, the dissecting-room; the nude living figure, surface and regional anatomy. Whatever the intent of the author may have been in designing this plate, we may see some traces of the organization in the text of the *Epitome*. The first chapter concerns the skeletal parts, the succeeding five chapters describe structures disclosed by dissection, and the book closes with a section on topographical and surface anatomy. The modern encyclopedias of anatomy follow almost exactly, chapter for chapter, the order in which Vesalius describes the systems of the body.

Another influence which may be observed in the organization of this book is that of Mondino de' Luzzi, whose *Anathomia* of 1316 starts with parts associated with digestion, the *natural members;* following this come the contents of the thorax, or *spiritual members,* and then the nervous system, the *animal members.* This order is exactly that of the *Epitome*'s Chapters III, IV, and V, and the general interpretations of these chapters will show the relationship to the three "spirits," which may be compared with the system of "members" of Mondino.

Immediately on beginning the reading of the *Epitome,* the question arises: for whom was this text intended? According to Vesalius, in his dedicatory introduction, it is a compendium of anatomy for physicians and may be treated as an index to the *Fabrica.* Singer, in *The Evolution of Anatomy,* believes that it was "intended for those who were not students of medicine" (p. 123). However, the first thing which arouses the curiosity of the reader is Vesalius' restraint in naming the parts described. Except for those structures the names of which are in common, even household, use, almost no names are given directly. Instead, the derivation of the anatomical name is often given by the device of descriptive comparison. In illustration, the "second cartilage" of the larynx "resembles the ring which the Thracians fit on their right thumb when shooting arrows" in spite of the fact that the term "cricoid" was already available (Gr., χρίχος, "a ring").

In this connection, a communication from Professor J. B. deC. M. Saunders of San Francisco informs me that in his experience with the *Fabrica* he has found the *Onomasticon* of Julius Pollux of great value in dealing with nomenclature. Singer (*op. cit.,* p. 107) says that this work, while written in the time of Galen, first became influential after its publication in 1502. Its section on anatomical terms became the storehouse from which the Humanists replaced the current Arabic terms. For example, "cricoid" was thus introduced.

The conclusion seems tenable that in the *Epitome* Vesalius has employed this picturesque device in the introduction of new terms much in the way that modern teachers do. We feel often that teaching is made more vital and the student's interest and retention of **fact** increased if we explain, as far as possible, the derivation of the term; we mourn the lack of training in the classical languages (both in teacher and in student) which prevents this from reaching its maximum utility. According to such an interpretation, the *Epitome* was written not for barber-surgeons and bonesetters, for whom sheer rote memorization of names and methods would serve, but instead for the medical student, for whom the greatest effort in offering a basic understanding of fact pays manifold dividends in the form of a well-trained physician.

Another teaching method introduced in the first chapter is enumeration of parts. Vesalius feels strongly that this is of value in anatomy teaching, as shown by the third paragraph of his dedication. It reaches its highest and most complex form in the second chapter on muscles; an attempt has been made to preserve its full force in the annotation. The method has been dropped, almost completely, in anatomy (an outstanding exception is the enumeration of twelve cranial nerves) but is in use in pathology teaching, where the student learns the four cardinal signs of inflammation, the five diseases which show leukopenia, the five types of cirrhosis of the liver, and other similar enumerations.

In connection with Vesalius' methods of teaching, it would be gross negligence to omit his manikin plates in the *Epitome*. We can imagine the great pride he took in this "first instance in which the type of demonstration by moveable layers is adopted" (Singer, *op. cit.,* p. 126). While the vascular system rather than the skeleton is the basis for this pair of male and female manikins, much care has been taken in designing the various portions of the plates so that they may be cut out and superimposed. The instructions on how to assemble them are written in great detail. Singer's comment that "It is not a useful method, and has since fallen into desuetude save for popular purposes" may be modified on the basis of the recent appearance of three volumes of this type, the first a detailed study of the body and the other two a series of anatomical transparencies of the eye and the ear. The use of transparent mediums in these demonstrations may yet bring about the reintroduction into anatomical teaching of a method invented by Vesalius.

NOTES

1. THE first few lines contain an idea which is now commonplace in elementary anatomy, viz.: organs are composed of tissues. Probably it was the first great generalization in the analysis and synthesis of the main anatomical facts.

2. Frontal bone. Vesalius appears to expect metopism (persistence of the metopic suture) but finds many skulls, especially female, which do not show it. Metopism is actually the exception, and significant sex differences are not now claimed for it.

3. Parietal and temporal bones.

4. Vesalius recognized the existence of the third auditory ossicle, the stapes, in his letter to Fallopius some twenty years later.

5. The zygomatic processes of the temporal and maxillary bones, yoked together by the zygomatic bone.

6. The petrous portion of the temporal bone, with special reference to the appearance of its inferior aspect.

7. The sphenoid bone, wedge-shaped or irregular in form (σφήν, "a wedge").

8. Compare the ancient name of this bone: sphecoid (from σφήξ, "a wasp").

9. The ethmoid bone, lamina cribrosa and lamina perpendicularis. The comparison to a sponge refers to the ethmoid air cells. See Chap. V, note 18.

10. "Upper jaw" here is the topographic equivalent of face, exclusive of mandible. The bones indicated are, laterally, the malar or zygomatic bone, and medially, the lacrimals and laminae papyraceae of the ethmoids.

11. Maxilla, with the alveoli.

12. Palatine bone.

13. Nasal bones.

14. Sinciput: *semi + caput,* the opposite of occiput; forehead or calvaria.

15. Lambdoid suture.

16. The squamous part of the temporal bone: *squamosae conglutinationes.* The term refers to a union of thin scalelike bones. See Morris' *Human Anatomy* (10th ed., 1942), p. 268.

17. Harmony: smooth union along a nearly straight line. This concept has continued to be the same to the present time. It reflects the Greek concept of fitness in joining or fastening, as in carpentry or masonry (and in anatomy), by mere apposition. Morris, *loc. cit.*: "In harmonic suture, the even, regular edges of the apposing bones. . . . "
18. Separate centers of ossification in the mandible and other bones; Vesalius' experience was not limited to the dissection of the adult body.

19. Mandibular fossa of the temporal bone and condyloid process of the mandible.

10

Sinuses here mean hollows, articular fossae. The word has long had many meanings in anatomy.

20. The hyoid bone; more curved than sharp at the convergence of its cornua. Vesalius would correct the old name of "lambdoid bone."

21. Trachea or windpipe; the idea is a derivative of the concept of arteries as air passages. "Rough" (Latin, *aspera*) is descriptive of the cartilaginous rings.

22. Thyroid cartilage (θυρεός, "a shield").

23. Cricoid cartilage (κρίκος, "a ring").

24. Old name, *cartilago innominata* (Fr., *cartilage anonyme*).

25. Arytenoid cartilages; *arytaena*, "ladle-shaped."

26. Corniculate cartilages and vocal folds (or cords). The diminutive of *cornu*, "little horn"; compare this with other terms applied, such as *tibia*, "flute," and *fistula*, "pipe." Both of these last appear to have been lost; they are not found in Motherby (*A New Medical Dictionary*, 2d ed., London, 1785), though they may be found as synonyms in Dunglison (*A Dictionary of Medical Science*, rev. ed., 1874).

27. Glottis or lingula: the mouthpiece of a pipe.

28. The epiglottis.

29. I.e., the uncial capital sigma.

30. *Medulla*, marrow, representing the concept that the white, fatty material filling bone cavities is comparable to that found in the cavity of the vertebral column.

31. From its fancied resemblance to the bill of a cuckoo (*cuculus*, κόκκυξ).

32. The atlas.

33. The epistropheus (axis), with the odontoid process (dens); odontoid means tooth-like.

34. Vesalius is here referring to the transitional vertebrae of the thorax. His qualification of "usually" indicates his full realization that this may be either the eleventh or twelfth thoracic vertebra. The remainder of the paragraph is doubtless a description of the differing inclinations of the various articular processes, but here, as all too frequently in the *Epitome*, the clarity of his concepts has been lost in overcondensation.

35. The sternum is "composed very rarely of seven bones"; a concession to Galen, who so described its composition.

36. I.e., the lower the rib, the shorter it is, and hence the less it extends toward the mid-line.

37. Xiphoid process. *Gladiolus* ("little sword") was the old name for sternum. Ensiform process carries the same meaning.

38. Manubrium, the "handle" of the sword.

39. The glenoid cavity.

40. *Summus humerus:* the acromion process. According to Professor Saunders, Galen described a third bone in the shoulder, the summus humerus, which Vesalius concluded was the acromion. There was a long dispute on this subject, and some concluded that the summus humerus was the cartilage of the acromioclavicular joint.

41. Coracoid process; other names indicative of its shape which have been in use are uncinate, corniculate, anchoralis, rostriformis, ancyroïdes. Cf. Vesalius, *Fabrica* I, xxi, on Galen's use of the term.

42. Including (as sinuses, see note 19) the coronoid, radial, and olecranon fossae, and (as tuberosities) the epicondyles, trochlea, and capitulum.

43. The olecranon process.

44. The articular disk.

45. On comparative anatomical and ossification bases, Vesalius holds the view, more recently advanced (and criticized), that the thumb has no metacarpal but is like the fingers in having three phalanges. Hence, the "four bones of the postbrachial" equal our second through fifth metacarpals.

46. Sesamoid bones. Nodes are joints; hence internodes correspond to phalanges.

47. Can Vesalius have referred here to the finding in arthritis (and especially ascribed to gout) of Heberden's nodes? The reason for the unusually large number of "sesamoid bones" found in connection with the other digits is not evident; there may be some confusion with epiphyseal ossification centers.

48. The limits of the three bones described here do not exactly correspond to those at present specified for the os coxae, but the principle of fusion of three bones is clearly defined. The "perforation" is the obturator foramen.

49. The description of this hinge joint is always one of the most difficult to accomplish with clarity because of its complexity; Vesalius is describing the method by which the two condyles and intercondylar fossa of the femur fit on the condyles and intercondylar eminence of the tibia, with the interposition of the cartilaginous menisci.

50. Greater trochanter (old name, *glutus,* "rump"); followed by lesser trochanter.

51. "Tibia" is not only a specific term for a single bone but also a generality for the leg region. This is explained also in the topographic or surface anatomy section at the end. The principle holds for several parts of the body, and the differentiation of bone or region is usually made from the presence or absence of the word *os*.

52. Calcaneum, which extends posteriorly beyond the perpendicular line formed by the posterior aspect of the tibia.

53. In the anatomy of the period, the tarsus is not so extensive as in modern terminology; it is limited to the navicular, the three cuneiforms, and the cuboid and excludes the calcaneum and talus.

54. *Ossa pedii* does not refer to all of the bones of the foot since the term is strictly limited to the metatarsal bones, in the same way that the "bones of the postbrachial" are always to be rendered as metacarpal bones in translation.

12

55. Note that he does not attempt to homologize the thumb and the great toe in particular, with respect to number of bones forming these digits (see note 45).

56. See note 46. The sesame seed was often symbolic in medieval metaphysics not only of minuteness but also of fertility and hence immortality. On this last meaning depends his next statement.

57. Surely Vesalius contributes here his own sardonic joke; he comments on the old superstition of the incorruptible bone, upon the foundation of which the whole body will be recreated on Resurrection Day. His training in philosophy must have been such that he would have appreciated the reference Butler made later to the os sacrum (holy bone, Vesalius' "meagre ossicle" of the coccyx) in his satirical poem:

> The learned Rabbins of the Jews
> Write there's a bone, which they call luez,
> I' the rump of man, of such a virtue,
> No force in nature can do hurt to. . . .
> —SAMUEL BUTLER, *Hudibras,* Part III, Canto II

Vesalius's transference of the property of indestructibility to an insignificant bone lying in a tendon under the big toe would have been adequate comment to answer (by anticipation) the questions of contemporary metaphysicists regarding his findings on incorruptible bones. The corresponding passage from the *Fabrica* will be found in translation in Clendening's *Source Book of Medical History,* p. 150. See Richard Burton's translation of *The Thousand and One Nights* (notes on the 449th night) and Dr. Nash's notes on *Hudibras* for interesting additional reading on this subject.

CONCERNING THE LIGAMENTS OF THE BONES

AND CARTILAGES, AND THE MUSCLES WHICH ARE THE
INSTRUMENTS OF VOLUNTARY MOVEMENTS

CHAPTER II

THE ligaments, no less than the sinews of the muscles and the organs by means of which the animal spirit is led from the brain, are commonly called nerves [1]. They are made of similar particles, originating from bone and cartilage [2], almost entirely devoid of sensation, hard (but nevertheless softer than the cartilage), and white; they serve various purposes in the fabric of man in binding together, containing, covering, and forming the muscles. For the muscle is regarded as the instrument of voluntary motion, formed of fibers interwoven with flesh and with many membranes having the nature of a ligament [3]. In order that the muscle may contract and lead the part that is to be moved [4], it requires the assistance of the nerves which carry the animal force from the brain no less than it requires the sense organs. Similarly, in order that the rest of the parts which require nutrients may be nourished, it is irrigated by veins and arteries. Further, the tendon of the muscle is a certain type of sinew and is, as it were, composed of fibers except for the intervening flesh. The tendon is distinct from the muscle only at the point where it reaches a position far enough from its origin (which it takes from a bone, cartilage, or some membrane) toward the insertion (which the tendon makes on the part that is to be moved) that its fibers attain continuous traction through the whole muscle; this is obtained by virtue of its interconnection with the proximal fleshy portion of the muscle. In proportion to their length, muscles end in tendons now round, now broad and membranous, now shorter, now longer; the short muscles spread out in a continuous expansion and terminate in no sinew worthy of mention.

A membrane augmented with fleshy fibers lies under the skin of the forehead. The skin is the natural outer covering of the body and sends out a thin little skin growing upon its own surface, like an efflorescence of the true skin itself [5]. Inwardly, throughout the entire body, a certain membrane lies under the skin; because it is increased with fleshy fibers (as here in the forehead), it has been thought to be called carneous elsewhere [6]. The largest part of a man's fat is placed between this membrane and the skin [7]. Also, the upper eyelid is moved by means of the membrane just mentioned, for in that place where this [carnea] reaches the inner region of the eye socket, it raises the eyelid [8]; where it is augmented at the external aspect of the eye by fleshy fibers drawn into the image of our letter C, it is the originator of the downward motion of the eyelid [9].

Seven muscles move the eye [10]: the first leads to the side inward, the second outward, the third upward, the fourth downward; the fifth and sixth are somewhat curved around the eye and render moderate assistance to the function of the first and second muscles. These six correspond to each other in form, being elongated and almost round; they spring forth from the tough membrane which surrounds the nerve of vision [11]. They pass by membranous tendons to the anterior portion of the scleral tunic of the eye in a ring near the border of the iris. The seventh muscle is covered by these and surrounds the visual nerve along with the six muscles at the same time [12]. The carneous muscle is implanted in the posterior portion of the scleral tunic of the eye; it is chief among those six muscles, making almost the same motions as they do.

A muscle draws one ala of the nose upward and outward, arising from the inner aspect of the cheekbone and inserting at one side into the ala, on the other into the upper lip where the latter lies below the ala of the nose [13]. A membranous muscle draws the alae together inwardly, lurking in the broad part of the nostrils beneath the tunic which girdles them from below [14].

Four muscles on either side move the cheeks and lips [15]. The first is composed of fleshy membrane reinforced for the most part with carneous fibers in the anterior portion of the neck and face as far as the cheeks; it resembles a muscle. The second arises from the cheeks and is implanted in the upper lip. The third extends from the lower jaw to the lower lip. The

15

fourth, somewhat different, lies in that part of the cheeks which we puff out. To these muscles there approaches a portion of that muscle which, as we learned, moves the outer ala of the nose; this, together with those mentioned, produces those marvelous and varied movements of the cheeks and lips.

There are also four muscles on either side which move the lower jaw [16]. The first is the temporal, arising from the bone of the vertex and also of the forehead, as well as from the bone shaped like a wedge; it is broad and flat at its origin from the bones of the temple and inserts in an acute process of the lower jaw. The second is called mansorius [17], from chewing [a mandendo]; it proceeds from that·part of the skull called the os jugale and is inserted in the external aspect of the jaw. The third comes from the processes of the skull which have the shape of wings; it is implanted in the internal aspect of the jaw; it raises the jaw, together with the muscles named, and moves the jaw laterally and forward and backward. The fourth, with its companion, depresses the jaw. It takes its origin from the process of the skull which resembles a stylus and inserts at the apex of the chin; it is furnished with two bellies for itself.

The hyoid bone [18] is drawn straight downward by two muscles, close to each other and arising from the top of the chest bone. Two others, coming from the lower jaw, draw this bone upward. At the sides it is drawn upward by one muscle on either side, originating from the styloid process of the skull. Having origin from the upper border of the scapula, one muscle at each side moves it downward. All eight of the above muscles are inserted in the anterior portion of the hyoid bone.

A fleshy mass, coming from the middle of the bone just mentioned, is inserted in the root of the tongue [19]; this mass can be regarded as two muscles and moves the tongue straight inward and downward. From the sides of the hyoid bone one muscle is also inserted into the root of the tongue inwardly but nevertheless moving the tongue more to the side. The fifth and sixth muscles of the tongue come from the styloid processes of the skull on either side; they are implanted in the root of the tongue and move it upward or, as the one or the other is contracted, to either side. The seventh and eighth muscles originate from the sides of the lower jaw near the root of the molar teeth on either side. They are inserted along the length of the tongue on the lower side and move downward and laterally that part of the tongue which is conspicuous in the gaping mouth before dissection. The ninth muscle arises from the inner aspect of the lower jaw near the tip of the chin; it is thick and has a few furrows or wrinkles. It is inserted in the lower part of the tongue and moves the tongue outward; its body as seen before dissection is interlaced with fibers of such a sort that by the great workings of Nature the tongue is very readily carried in every direction of movement [20].

Four muscles contracting the little crevice of the larynx join the first cartilage of the larynx to the second. Four muscles opening this crevice bind the third cartilage to the second. Two closing the little crevice connect the third cartilage to the first. Two others set in the base of the third cartilage contract the crevice tightly. These twelve muscles are said to belong to the larynx [21]. Two additional ones proceed together from the hyoid bone and insert into the first cartilage; they elevate it in front and also retract the crevice of the larynx [22]. Two muscles arising from the breastbone extend to the same cartilage [23]. Then two arising in turn from the posterior region of the esophagus (completely fleshy, as are almost all of the laryngeal muscles) are also inserted into the sides of that cartilage [24], together with the two already mentioned as contracting the larynx. Two others extend from the hyoid bone and insert into the root of the epiglottis; they pull the latter up and forward [25].

In the group of muscles which move the head are also those which move the first cervical vertebra separately [26]. There are seven pairs of them in all, because the same number, of course, is reckoned on each side. The first pair passes from the spines of the upper five thoracic vertebrae; ascending gradually in an oblique direction outward, they are implanted on the occipital bone. The second pair, apparently formed not of two but of a number of muscles, is relatively different; a special portion of it arises from the transverse processes of the four upper thoracic and five lower cervical vertebrae and, passing inward somewhat obliquely, inserts on the occipital bone. The third pair arises from the spine of the second cervical vertebra and, creeping outward somewhat obliquely, also inserts on the occipital bone. The fourth pair, likewise inserted on the occipital bone, passes from the first vertebra at the place which, in other vertebrae, ends in a spine. The fifth pair is carried somewhat transversely from the middle of the occipital bone to the lateral processes of the first vertebra. The sixth pair extends from the spine of the second vertebra to the same processes; like the five pairs just mentioned, it is located in the posterior region of the neck; like the third, fourth, and fifth pairs, it is composed of muscles which are completely fleshy, round, and slender. The seventh pair is more noteworthy; from the top of the breastbone where the clavicles articulate, two muscles (one on each side) have their origin and, borne obliquely upward, they extend to an insertion on the mastoid process of the head. When the first four pairs are tensed at the same time, the head is moved back; when the muscles of the first three pairs are tensed from the opposite side they assist in turning the head, and in that movement the fifth and sixth pair turn the first cervical vertebra and the head circularly. The function of the seventh pair of muscles, operating together, is to direct the head straight forward, but when they work alternately, they turn the head around. The muscles of the neck subserve the motion of the head to the extent that the latter is turned with the neck secondarily and is led laterally toward the shoulders [27].

Among the eight pairs of muscles which move the back, these muscles are to be classed [28]. The first pair, beginning from the sides of the fifth thoracic vertebra, extends to the first cervical vertebra below the esophagus and flexes the upper part of the back. The second extends on either side from the first rib of the thorax and inserts on the inner portion of the trans-verse processes of the cervical vertebrae; it turns the neck to the side but rather more in front. The third, taking its origin from the transverse proc-esses of the upper six thoracic vertebrae, is implanted in the external region of the transverse processes of the cervical vertebrae; it leads the neck back-ward, inclining the neck to the side. The fourth pair extends from the spine of the seventh thoracic vertebra to the second cervical vertebra, inserting into all intervening vertebrae, as does the first pair, and originating from them also; it extends the upper part of the back. The fifth pair has a muscle on either side from the iliac bone to the transverse processes of the lumbar vertebrae and inserts on the lowest rib of the thorax; it flexes the lower part of the back. The sixth, arising in the back from the lowest aspect of the os sacrum, extends to the neck; it inserts on the transverse processes of the lumbar vertebrae and even more distinctly on those of the thorax. When both muscles of this pair contract, the back is extended; when only one of the pair is in action (as among the other pairs also), this provides an oblique movement or toward the sides. The seventh, hidden beneath the sixth, originates from the posterior aspect of the os sacrum; attached to all the intermediate spines, it ascends as far as the spine of the eleventh thoracic vertebra and, by successively drawing the spines together, it extends the back in this region. The eighth pair, extending from the eleventh thoracic vertebra to the seventh cervical vertebra, is completely joined to the inter‑ vening spines in the same fashion as is the seventh.

One muscle moves the scapula to the chest [29]; it arises from the second, third, fourth, and fifth ribs of the thorax before their cartilaginous termina‑ tion and is inserted in triangular fashion on the inner process of the scapula. A second muscle of those which move the scapula arises from the occiput and, following along the length of the neck to the spine of the eighth thoracic vertebra, originates from the apexes of the vertebral spines as well; it inserts on the spine of the scapula, the acromion, and a portion of the clavicle. All of that part of it lying in the neck pulls the scapula upward; that part below the neck in the posterior portion of the thorax and which resembles a monk's hood, draws the scapula downward. The third springs from the transverse processes of the upper cervical vertebrae; this muscle inserts into the upper angle of the base of the scapula and lifts it upward. The fourth arises from the spines of the fifth, sixth, and seventh cervical vertebrae and especially from the spines of the first three thoracic vertebrae; this muscle inserts at the base of the scapula and draws it somewhat backward.

The first author of the movements of the arm [30] arises from the middle region of the clavicle nearest the breastbone and from the breastbone; narrowed into somewhat of an angle, it draws the arm to the chest. The second muscle, originating from the other part of the clavicle, the point of the shoulder, and from the spine of the scapula, is inserted transversely by its apex into the humerus; beautifully covering the shoulder joint, it lifts the arm upward in various directions. Its shape is not unlike that of the

19

Greek letter Δ. The third muscle proceeds from the lower border of the scapula; it draws the arm straight backward. The fourth muscle takes its origin from the spine of the sixth thoracic vertebra to the lower aspect of the os sacrum and from the apexes of the intermediate spines. It is drawn together into the vertex of a triangle and is inserted into the humerus where the three muscles just enumerated insert, far below the articulation of the head of the bone with the scapula. This muscle draws the arm downward in various directions, since its base is quite broad and not led from a point in such a fashion that the muscle would exercise a wholly simple movement. The fifth muscle occupies the entire cavity of the scapula facing the ribs. The sixth appropriates to itself the entire gibbous part of the scapula lying below the spine. The seventh fills the conspicuous sinus between the spine of the scapula and its upper border. These three are inserted by a broad implantation into the ligaments surrounding the shoulder joint; they accomplish the rotation of the arm [31], and the seventh seems to be of some assistance in raising the arm.

The first muscle of those which move the thorax [32] originates from the clavicle and is inserted into the first rib of the thorax; it moves this rib upward, at the same time assisting the dilatation of the thorax. The second arises from the base of the scapula and is inserted with digitations into the upper eight ribs long before they terminate in cartilage; it moves the ribs outward and dilates the thorax. The third muscle takes its origin as a broad membrane from the apexes of the three lower cervical spines as well as from the first thoracic vertebra and inserts with three digitations into the three interspaces of the four upper ribs beneath the base of the scapula; it moves these ribs obliquely upward and enlarges the thorax. The fourth originates on the iliac bone and passes upward to the neck, it is inserted into the twelve ribs where they first leave the vertebrae, and it contracts the thorax. The fifth originates from the apexes of the spines of the two lowest thoracic and some of the lumbar vertebrae. It is membranous, passes transversely, and is inserted into the ninth, tenth, and eleventh ribs at the place where they turn back into the deep parts; it dilates the thorax. The sixth lies back in the breadth of the thorax; it extends to the cartilages of the true ribs and to the side of the breastbone and contracts the thorax.

Internal and external muscles lie in the intervals of the twelve ribs [33]. Those external ones which are in the intervals between the bony ribs send their fibers from the upper rib obliquely to the lower rib straight opposite; the internal muscles send their fibers from the lower rib obliquely upward in the anterior direction to the upper rib. In the six intervals of the cartilages which are allotted to the true ribs, the fibers of the external muscles creep from the lower cartilage to the opposite upper cartilage in an oblique direction, but the internal fibers extend in reverse from the upper cartilage to the lower.

Hence, muscles in groups of four are counted in the intervals of the six true ribs. In the interspaces of the false ribs, however, there are groups of only two. All the intercostal muscles on one side total thirty-four; all of them have the function of contracting the thorax. Forty muscles are enumerated thus far on one side of the thorax; there are the same number on the other side, and to these eighty muscles one is added, to wit, the septum transversum [diaphragm] itself [34], inserted into the lowest part of the breastbone and the cartilages of the false ribs as well as into the upper lumbar vertebrae. In the middle it is sinewy [35] but fleshy circumferentially toward the insertion; it divides those organs which serve for making blood and for generation from the region of the heart and of the parts which minister to the heart [36]. It has the function of dilating the thorax.

To these are added eight muscles of the abdomen [37], four on each side. The first or outermost sends its fibers obliquely downward and forward, forming with its mate a covering for the abdomen. The second sends its fibers obliquely upward in the opposite direction and with its mate forms a covering for the abdomen. The third sends its fibers straight upward from the pubic bone to the chest. The fourth has its fibers distributed transversely and with its mate forms a covering for the abdomen, as do the oblique muscles; it renders assistance, no less than the other abdominal muscles, to the constriction of the thorax [38].

Two muscles flex the elbow [39], of which the anterior derives one head from the higher region of the neck of the scapula and another head from the internal process of the scapula; formed with these heads, it is inserted into the radius. The posterior muscle originates from the humerus and is inserted into the anterior region of the elbow joint or, rather, into the ulna. Three muscles extend the elbow [40]; one originates from the lower border of the scapula, and the second from the posterior aspect of the neck of the humerus. These merge together in their descent and a third joins with them, arising from near the middle of the length of the humerus and inserting together with them into the posterior process of the ulna.

In the inner aspect of the elbow lies a slender muscle which arises from the inner protuberance of the humerus and turns into a flat tendon lying beneath the internal skin of the greater part of the hand. By its function it is believed that this skin is rendered less movable and more fitted for grasping [41].

The radius is led into pronation by two muscles [42]. One arises from the inner region of the elbow joint and is implanted obliquely on the radius; the other is borne from the ulna transversely to the radius near the wrist. The radius is led into supination [43] by two other muscles, one of them long and extending from the humerus to the lower part of the radius, to which the wrist is articulated. The other extends obliquely from the outer aspect of the elbow joint to the middle of the length of the radius and inserts there.

The wrist is moved by four special muscles [44]; the first two grow forth from the inner protuberance of the humerus. One is inserted in the postbrachial bone which supports the index finger [second metacarpal], the other on the smallest bone of the wrist [pisiform]. The third arises from the humerus and is inserted with a bifid tendon into the postbrachial bones which sustain the index and middle fingers. The fourth, passing from the external tuberosity of the humerus and extending along the ulna, forms an insertion on the postbrachial bone which supports the little finger. The first two simultaneously flex the wrist; when the third and fourth are simul-taneously contracted, they extend it. When the first is tensed with the third, the wrist is moved toward the inner side; when the second and fourth act together, it is inclined to the outer side.

The first of the muscles which move the fingers of the hand [45] arises from the inner and anterior region of the elbow joint; proximal to the root of the wrist it splits into four tendons, inserted into the second internodes of the four fingers and flexing the internodes. The second arises from the same place as the first but is more slanting and lies under the first muscle. It is also divided into four tendons, lying under those of the first; they pierce those tendons proximal to the root of the second internode of the fingers. Finally they form an insertion in the third bones of the four fingers and flex them. The third arises from the radius near the elbow joint, is inserted into the third joint of the thumb, and flexes it. Thirteen other muscles follow the third muscle in series; they lie in the hand, and of them, two are inserted into the first bone of each of the five fingers and flex them. Three insert into the second internode [first phalanx] of the thumb also and move it. The seventeenth muscle of those which move the fingers arises from the external protuberance of the armbone; it is inserted in the index, middle, and ring fingers especially and extends those fingers. The eighteenth muscle proceeds from the same region as the muscle just mentioned; it is the prime originator of the extension of the little finger and, blended to a varying extent with that tendon of the seventeenth which is inserted into the three bones of the ring finger, it subserves therein the abduction of that finger toward the exterior. The nineteenth, along with the twenty-first, has a common origin from the ulna; it arises near the region of the wrist. Almost divided into two tendons, it sends one to the outer side of the index, the other to the side of the middle finger, and is considered the author of the abduction of those fingers to the external side. The twentieth arises from the wrist; it extends along the external side of the postbrachial bone which supports the little finger, inserts on the first bone of that finger, and moves it outward to the side. The twenty-first is inserted on the outer side of the thumb as far as the third joint; it extends the thumb toward the index finger. The twenty-second proceeds from the ulna a little above the muscle just mentioned, and soon splits into two parts, one part ends in a tendon; it inserts on the bone of the wrist which supports the thumb and, passing to the place where the hand follows the motion of the radius in pronation, it lends assistance to that motion. The other part is likewise split into two parts which form one tendon each. Of these, the one inserts into the external side of the outer region of the first bone of the thumb; the other grows on that bone and is inserted on the second and third bones of the thumb. By

23

the function of these tendons, the thumb is flexed inward. The twenty-third occupies the region near the inner side of the first bone of the thumb and distinctly separates the thumb from the index finger. The twenty-fourth springs from the bone of the postbrachial which supports the index finger, is inserted chiefly into the first bone of the thumb, and brings the thumb close to the index. There remain four slender muscles extending in the palm from the four tendons of the second muscle of those which move the fingers. They are inserted on the inner side of the first bone of the four fingers, serving the abduction of those fingers thumbward.

On the internal aspect of the elbow lies the muscle which forms the broad tendon of the hand: the first and second muscles which cause the movements of the wrist; the first, second, and third of those which move the fingers; and two muscles which pronate the radius. In the external aspect lie the seventeenth, eighteenth, nineteenth, twenty-first, and twenty-second of the muscles which move the fingers; the third and fourth of those which have charge of the movements of the wrist; and two which supinate the radius. These total nine; there are ten if you distinguish from the twenty-second of those muscles which move the fingers that portion which offers a tendon to the bone of the wrist which supports the thumb. In the hand itself ten muscles are observed which flex the first joints of the fingers; three flex the second internode of the thumb. Then there are the twentieth, twenty-third, and twenty-fourth of those which move the fingers, and four muscles by which the four fingers are moved toward the thumb.

Each of the male testes, with their seminal vessels, is covered by a tunic which proceeds from the peritoneum [46] and is nourished by some straight fleshy fibers [47] inserted into the lowest region of the vessel carrying the semen [48]. One muscle of the testis is made up of these fibers; by its function the testis is drawn upward closely. Thus also the membranes which secure the uterus are equipped on either side with fleshy fibers, and in this way the uterus has one muscle on either side by the assistance of which it is easily drawn upward toward the ilia [49]. One muscle, preventing the untimely excretion of the urine, arises circularly on the neck of the bladder [50]. Likewise, there is also a muscle which encircles the end of the rectum;

it prevents premature expulsion [51]. Two other muscles pull the rectum quickly upward after expulsion [52]. A slender muscle is inserted at the root of the penis from the pubic bone on either side, assisting its erection very slowly [53]. Likewise, two muscles arise close to each other from the anterior region of the muscle which embraces the rectum circularly; they are inserted at the urinary passage where it turns back and upward under the bones of the pubis. They dilate the passage during the ejaculation of semen so that it may not be shut off at the bend [54].

The first of the muscles which move the thigh [55] arises from the outermost aspect of the iliac bone and the posterior region of the coccyx; it passes to the posterior region of the greater process of the femur and also is attached by a broad insertion into its root. The second muscle is hidden under the first, in large part, but extends rather more forward from the anterior region of the iliac bone and is inserted into the greater process of the femur. The third is much smaller than the second muscle and is completely hidden by it. It arises from the iliac bone near the posterior region of the acetabulum of the coxendix [hipbone]; it is also inserted on the greater process of the femur. Like the two previously mentioned, it extends the femur, moving the latter outward to the side. The fourth muscle extends from the three lower bones of the os sacrum, also inserts on the greater process, and extends the femur and turns it outward to some extent. The fifth is the largest of all the muscles of the body and, with many parts, it takes its origin from the bone of the coxendix [ischium] and of the pubis and is inserted into the posterior region of the femur as far as its lower heads [condyles]. This muscle is the author of extension of the femur, holding it upright and moving it inward with a portion of it especially thrust forward from the lower region of the pubic bone. The sixth muscle takes its origin from the two lowest thoracic vertebrae and from some of the higher lumbar vertebrae; inserted into the lesser process of the femur, together with the seventh muscle, it flexes the thigh. The seventh muscle proceeds from the entire internal aspect of the iliac bone; it also is inserted on the lesser process, higher than the sixth. The eighth passes from the pubic bone and is implanted with a long insertion below the lesser process of the femur; it flexes the thigh and also moves the thigh strongly inward. The ninth muscle occupies the anterior region of the foramen of the pubic bone; inserted into the greater process of the femur, it turns the femur inward. The tenth occupies the posterior or internal aspect of the foramen just mentioned and is very securely bent around the posterior part of the coxendix bone [ischium]. Like the other muscles here, it is increased by muscles arising from it. It is inserted on the greater process of the femur and turns the thigh outwardly.

The first muscle of those which move the leg [56] proceeds from the anterior region of the spine of the iliac bone and passes somewhat obliquely along the internal region of the thigh; it is inserted into the anterior region of the tibia and is at once the most slender and the longest muscle of the entire body. The second muscle proceeds from the union of the pubic bones: it inserts into the same region as the first. The third muscle begins from the appendix of the coxendix bone [ischium]; it is also implanted in the same region of the tibia. The fourth muscle passes forward from the same region of the coxendix bone and in its descent receives a portion of its sub-stance from the bone of the femur; it is inserted into the articulation of the tibia with the fibula, but especially into the fibula. The fifth muscle originates also in the same region and is inserted on the anterior aspect of the tibia with the first three muscles but in a less slanting direction. The sixth grows out from the spine of the ilium; it is covered by a sort of membrane along with the muscles investing the femur and is inserted in the knee joint rather near the outer side. The seventh muscle originates from the root of the greater process of the femur and occupies the external side of the thigh; it forms a tendon with the eighth and ninth, to which the patella is attached. The eighth arises from the neck of the femur and from the base of the latter's greater process; this muscle closely encircles almost all of the femur. The ninth, taking origin conspicuously in the anterior region, arises from the pro-tuberance of the hipbone above its joint with the femur; lying on the seventh and eighth muscles, the ninth muscle is carried to the anterior region of the knee. Implanted very firmly on the anterior règion of the tibia, this muscle turns into a tendon, forming one with the two muscles just men-tioned. Thus the first, sixth, seventh, eighth, and ninth muscles are con-

26

sidered the agents of the extension of the tibia, while the second, third, fourth, and fifth clearly flex the tibia.*

The muscle which is hidden in the popliteal region [57] and which extends obliquely from the external ligament of the knee joint to the tibia, does not flex the leg; if it does anything, it vaguely imitates the motion of the first muscle which pronates the radius.

Of the muscles which move the foot [58], the first originates from the inner head [condyle] of the femur near the knee joint, as the second begins from the outer head; both form the posterior part of the calf of the leg, and joining with the tendon of the fourth muscle which moves the foot, they are inserted on the heel. The third is a small muscle also proceeding from the external head of the femur and here in the popliteal region terminates in a very slender tendon which is inserted into the inner side of the os calcis. The fourth, the largest of those which move the foot, begins from the articulation of the fibula with the tibia; it ends in a very strong tendon which is united with the tendon of the first two muscles likewise inserted, together with it, into the heel. The fifth muscle is placed very deeply in the posterior region of the tibia and the fibula. It originates from those same bones where they first separate, and near the rear of the internal aspect of the malleolus, it sends forth a tendon to a bone of the tarsus; the tendon is inserted in this bone, which is contiguous to the bone resembling a die [59]. The sixth, proceeding from the tibia where the fibula is articulated with it superiorly, is situated in the anterior region of the leg; its tendon inserts on the root of the metatarsal [os pedii] which supports the great toe. The seventh arises from the fibula, occupies the external side of the leg, and inserts a tendon, reflected under the inferior part of the foot, on the bone which supports the great toe. Inserting a tendon on the root of the bone which supports the little toe, the eighth is hidden under the seventh and also arises from the fibula. The ninth is part of that one which, as I am about to describe, extends the toes of the foot; it is inserted at almost the

* . . . . atque ita primus, sextus, septimus, octavus et nonus tibiae extensionis opifices habentur, secundo interim, tertio, quarto et quinto tibiam liquide extendentibus is the reading in the Latin text.
To avoid the repetition of idea in extensionis and extendentibus which makes nonsense of this passage as it stands, I am forced to conclude that Vesalius meant to write flectentibus instead of extendentibus. I therefore emend the passage to read flectentibus and translate accordingly.

middle of the length of the metatarsal supporting the little toe. The foot is extended with the first five muscles or is firmly set upon the ground, although the third muscle performs this function feebly and, if it contributes at all to the movement of the foot, it turns the foot obliquely inward. The foot is flexed by the sixth, seventh, eighth, and ninth muscles, and by their function it makes lateral movements, in so far as each muscle moves it.

The first of the muscles which move the toes [60] is located entirely in the sole of the foot and in its lowest part possesses an exceedingly thick membrane acting as an intimate covering not unlike the broad tendon in the hand. This muscle arises from the bone of the heel and sends single tendons to the second internodes of the four toes; these tendons flex the toes. The second and third muscles creep through the posterior region of the leg; the second arises from the tibia more than the fibula and, having arisen from it, sends into the sole of the foot a tendon which then divides into four tendons. One of these is inserted into the third bone of each of the four toes; the tendons, as happens also in the hand, pierce the tendons of the first muscle and flex those bones. The third arises from the joint of the fibula to the tibia and, extending more from the fibula, it sends a tendon to the sole of the foot, whence a small portion of it mingles with tendons which flex the third internode of the index and middle toes. The remainder of the muscle is inserted as a whole into the second bone of the great toe and flexes it. Ten muscles, mutually intermingled in a remarkable way, succeed these; attached to the metacarpals, they flex the first bones of the toes, with two muscles extending to each of the toes. The fourteenth muscle, of which the ninth muscle of those which move the foot was reckoned a part, arises from the anterior region of the tibia and is divided into four tendons inserted into the four toes and regarded as the cause of their extension. The fifteenth also proceeds from the anterior region of the tibia; this muscle inserts on the great toe and is the master of this toe's extension. The sixteenth muscle lies in the upper region of the foot and is a fleshy mass divided into four tendons; of these, one is inserted on the external side of the upper aspect of the great toe, the second in that of the index, the third in the middle toe, and the fourth on the annular digit. These tendons move these toes toward the outer side. The seventeenth muscle occupies the external side of the foot and is inserted on the first bone of the little toe; it moves this toe away from the other toes. The eighteenth muscle extends from the internal side of the foot; it abducts the big toe straight from the others. Then in the sole of the foot is a fleshy substance [61] divided into four slender portions which cling to the tendons by the function of which the third bones of the four toes are flexed. These portions are inserted on the inner side of the four toes at the first joint; they are considered the authors of the adduction of the toes toward the great toe.

If you will count this last as four portions instead of four muscles, you will observe in the posterior region of the leg the first, second, third, and fourth of those muscles which move the foot, the second and third of those which move the toes, and under these the fifth of those which move the foot. In the anterior region lie the sixth, seventh, eighth, and ninth muscles which cause the movements of the foot and the fourteenth and fifteenth of those which move the toes. In the foot you will have the first of those muscles which move the toes and the ten muscles which flex the first bones of the toes and the sixteenth, seventeenth, and eighteenth also of those which move the toes, unless it be proposed to divide the sixteenth into more than one muscle.

In the description of the muscles, I have not made mention at all points of the ligaments, since for the most part they are mutually related with the joints. In all the joints is a ligament [62] running circularly from one bone to another or to a cartilage or from a cartilage to a bone, or the cartilage insertion extends from it separately. Special ligaments are an accession to a few joints. In the joint of the skull is a certain smooth round ligament running from the dens of the second cervical vertebra to the occipital bone [63]. Along the posterior region of the dens, another one is carried transversely in the first vertebra [64]. The bodies of the vertebrae are joined by somewhat cartilaginous ligaments [65]; the ascending and descending processes of the same bodies are connected with strong ligaments, but these only go roundabout in circular fashion [66]. Then a membranous ligament is situated in the intervals of the spines, as also in the forearm and the leg, where the bones are separated from one another [67]. In addition, in the foramen of the pubic bone occurs a ligament, or rather a membrane, of this nature [68]. In the joint of the humerus, three special ligaments are observed [69]. The first is round, arises from the internal process of the scapula, and runs to the external aspect of the head of the humerus. Two others arise from the higher region of the neck of the scapula; they extend to the same head, and one also is here led from the internal process of the scapula to the acromion [*summus humerus*]. In the joining of the bones of the wrist among themselves and with the metacarpal bones, as also in the foot, cartilaginous ligaments intervene. From the os sacrum two roundish ligaments extend to the ischium [70]. From the superior part of the head of the femur, a round ligament is inserted into the acetabulum of the hipbone [71]. A cartilaginous ligament is in the middle of the knee joint, and in the posterior region of that joint and on both sides, a special ligament is obvious to dissectors [72]. In the number of the ligaments covering the tendons transversely and containing the tendons, lest they slip from their seat [73], there is one in the inner region of the wrist which may be considered one and continuous, following the entire inner aspect of each digit [74]; six occur next to the root of the wrist in the outer aspect of the radius and ulna [75]. A ligament is also observed in the anterior region of the leg near the talus [76], and three between the heel and the internal malleolus [77]. There is one between the heel and the external malleolus [78]. Also, ligaments of this nature are observed in the internal or inferior aspect of the toes.

GENERAL INTERPRETATION

IN attempting the identification of the muscles described in this chapter, a technique of annotation has been adopted which is well illustrated by note 16. The identifying name of the muscle group given in the text is stated briefly, and the numbers following correspond to the order in which the muscles themselves are described. The method serves the three following purposes:

1. By reducing the number of footnote reference marks, the continuity of the text is less disturbed.
2. Since, in subsequent chapters, the text makes several references to muscles by statements such as "the third muscle of those which move the jaw" and since the legends for the muscle plates are in similar form, it is intended that more rapid identification of the structure to which reference is made may result. It should be stated here that a complete collation of notes and plate legends has not been possible but that in a number of trials the device has proved useful.
3. Emphasis is placed where the text places it, on muscle groups which act on, and across, joints. The text's discussion of individual muscles is minimized; instead of directing attention to the minute details of their bony relationships, they are treated as functional masses. An increasing number of anatomists at present is recognizing the soundness of this view not only in kinesiology but as a basic philosophy of human biology. Yet this is only a rediscovery of a principle derivable from Vesalius' text and stated more than two centuries earlier with conciseness and clarity by Mondino: " . . . . we must first gain an idea of the whole, and then of the parts. For all our knowledge doth begin from what is known. For though the known is oft vague and though our knowledge of the whole is of a surety vaguer than that of the parts, we yet begin with a general consideration of the whole" (Charles Singer, "Fasciculo di Medicina," *Monumenta Medica,* series Vol. II, Part I, Florence: R. Lier and Co., 1925, p. 59).

While the outline of Vesalius' text corresponds to the "systematic" anatomy, the chapters themselves are written as functional anatomy (or, to borrow a term noticed recently, "dynamic morphology"). Vesalius' conception of muscles as contractile masses associated with a joint and acting across it in co-ordination is more in accord with our ideas of synergy than is the treatment of them as single structures. The reasoning which derives a muscle's action by considering it as an isolated contractile element tending to shorten the distance between its points of attachment is incorrect physiologically (for it is rare, except in pathologic states of major paralysis, for a muscle to act alone); it is incorrect embryologically and neurologically; finally, it has served to introduce errors which have been carried forward through many editions of the standard encyclopedias of anatomy. The modern student has difficulty in finding in any of these great reference texts as clear a concept of integration as was available to the Renaissance student who owned a Vesalian *Epitome;* only with effort do we lead the student from the morass of isolated osteologic and myologic textual statements to the terra firma of synergy.

We are now being offered texts which are regional and functional rather than systematic anatomies. Some teachers are becoming more outspoken in their criticism of the teaching of anatomy as a discipline in morphology alone. To select one from many, a recent discussion of the approach to the teaching of anatomy states that "Stress is laid on the movements at joints and on their limitations. . . . The muscles are considered in functional groups rather than as separate units. . . . In this department we have tended to neglect . . . the memorizing of muscular attachments in relation to a bone instead of to a joint. . . . " (Sheehan, "The Physiology of Anatomy," *J. A. Am. M. Coll.*, XV [1940] 363).

The whole field of integration of muscle group activity is being subjected to the laboratory investigation it so badly needs, and we may anticipate in the not so distant future the time when all anatomists take up where Vesalius left off with the teaching of this aspect of functional anatomy.

NOTES

1. VESALIUS follows the old view which confused ligaments and nerves, or regarded the terms as synonymous. The Greek word from which comes our "nerve" originally meant *tendon* or *sinew;* hence the confusion. Observe (*a*) the relic of this usage which remains to us in the term "aponeurosis"—the flattened and membranous type of fibrous muscular attachment as opposed to "tendon" (example: the *aponeurosis* of the external abdominal oblique muscle, the *galea aponeurotica* of the epicranium); (*b*) the students' nickname for the slender, shining long tendon of the plantaris muscle: "Freshman's nerve."

2. "Originating from": *principium ducens*—not in the histogenetic sense of originating, but "leading forth from," in the modern anatomical sense of a muscle's origin.

3. Tendinous membranes to which are attached the muscle fibers, as in a multipennate muscle such as the deltoid.

4. Lead: compare the modern colloquial name for tendons, "leaders."

5. Dermis and epidermis.

6. The deep fascia; the extended argument is presented in the *Fabrica,* and Vesalius does not consider it to be carneous.

7. The superficial fascia, with its panniculus adiposus.

8. Levator palpebrae superioris muscle.

9. Orbicularis oculi muscle. The meaning of the paragraph is thus: where the connective tissue of the tendon attached to the upper border of the superior tarsus (and the fascia-sheath overlying this tendon) is followed back into the orbit, the fleshy (muscular) portion encountered is the levator; however, outside the palpebral fascia may be seen the incomplete circle of the sphincterlike muscle opposed to the levator —the orbicularis oculi.

10. Muscles which move the eye: (1) rectus medialis, (2) rectus lateralis, (3) rectus superior, (4) rectus inferior, (5) and (6) obliqui superior et inferior.

11. The fibrous ring annulus tendineus communis and the optic nerve. It is, of course, incorrect to include the inferior oblique as among the muscles arising from the annulus, for the former's origin is much more anterior, just lateral to the lacrimal sulcus. It is thus more equatorial in direction as compared to the meridional orientation of the first five named.

12. The seventh muscle of the eye is the retractor oculi (or *choanoides,* "funnel-shaped"), not present in human beings. In the sheep's eye it may be dissected out as a sheet of muscle underlying the other six, incompletely divided longitudinally, attaching into the sclera behind the recti, and passing back to surround the optic nerve. See Sisson and Grossman, *The Anatomy of Domestic Animals* (1938), p. 883, for discussion of this muscle. Cf. *Fabrica* XII, xi; also Cushing's *Bio-Bibliography,* pp. 186

and 190 for the exchange of opinion between Fallopius' *Observationes Anatomicae* and Vesalius' *Examen* (1561).

13. Quadratus labii superioris: angular head on the one (medial) side, and infraorbital and zygomatic heads on the other (lateral) side.

14. Nasalis, especially its alar portion.

15. Muscles which move the cheeks and lips: (1) platysma, including the risorius, which is not always well separated; (2) zygomaticus; (3) mentalis, quadratus labii inferioris, and triangularis, treated as a single mass; (4) buccinator—observe the correctness of its inclusion with the other muscles now known to arise from the second branchial arch and innervated by the facial nerve. The origin of the name is from *buccinare*: "to sound the trumpet"; compare Vesalius' observation on "inflate" (the cheeks). The muscle is set aside as "somewhat different," as if it is recognized that it is less a mimetic muscle than one associated with the next group, the muscles of mastication.

16. Muscles which move the jaw: (1) temporalis, (2) masseter, (3) pterygoidei, (4) digastricus.

17. Dunglison (*A Dictionary of Medical Science*, rev. ed., 1874) gives *buccinator* as the synonym. However, the muscle referred to is obvious from attachments given in the text, the buccinator has already been described, and Motherby (*A New Medical Dictionary*, London, 1785) indicates the masseter, as also Cooper, *et al.* The name "mansorius" is not found in Hyrtl's *Onomatologia Anatomica,* usually a rich source of anatomical synonyms.

18. Muscles moving the hyoid bone: (1) sternohyoids, (2) geniohyoids, (3) stylohyoids, (4) omohyoids.

19. Extrinsic muscles of the tongue, in pairs: (1) aponeurotic fibers of genioglossus and hyothyroid membrane; (2) hyoglosus; (5 and 6, the third pair) styloglossus; (7 and 8, the fourth pair) mylohyoids (which do not actually insert into the tongue but form a raphe inserting on the hyoid bone and make a diaphragm on which the tongue rests). The "ninth muscle" is the genioglossus.

20. The intrinsic muscles (generally oriented in three directions: vertical, longitudinal, and transverse).

21. Muscles belonging to the larynx: first four are the cricothyreoidei (pars obliqua et pars recta) paired, second four are the cricoarytaenoidei laterales et posteriores, next pair is the thyreoarytaenoidei, and the last pair the arytaenoidei. This makes the total of twelve (six pairs).

22. Thyrohyoidei.

23. Sternothyroidei.

24. Constrictor pharyngis, probably inferior.

25. Hyoepiglottic ligament.

26. Muscles which move the head and first cervical vertebra: (1) splenius capitis et

cervicis, (2) probably semispinalis capitis, (3) rectus capitis posterior major, (4) rectus capitis posterior minor, (5) obliquus capitis superior, (6) obliquus capitis inferior, (7) sternocleidomastoidei.

Observe that Vesalius probably turned the cadaver after demonstrating the hyoid and laryngeal musculature, since all of the pairs above except the seventh are most easily approached posteriorly.

27. I.e., indirectly rotate the head by rotation of the neck and do not directly act on the head.

28. Muscles which move the back: (1) longus colli, (2) scalenus anterior (and possibly including scalenus medius; these muscles vary in attachment and arrangement of fibers), (3) longissimus cervicis (portion of sacrospinalis), (4) semispinalis cervicis, (5) quadratus lumborum, (6) longissimus position of sacrospinalis, (7) multifidus, especially the more prominently developed lumbar portion, (8) semispinalis dorsi. The meaning of the last sentence is that these last two named are not uninterrupted bands of muscle but composed of many slips, with attachment to all vertebrae intervening between their termini.

29. Muscles which move the scapula: (1) pectoralis minor, (2) trapezius, (3) levator scapulae, (4) rhomboidei major et minor. The monk's hood is an apt comparison for the trapezius; some readers may be more familiar with the shape of the (derived) hood worn with academic dress and may carry the likeness forward from that. The text reads as if the trapezius functions either to elevate or to depress the scapula; its mechanism is not always clearly defined today. Actually, it probably *rotates* the bone, the upper part elevating the outer portion of the scapular spine, the lower (thoracic) part simultaneously depressing the inner portion of the spine. The center of rotation is in the spine and receives the predominantly aponeurotic, nonmuscular portions of the trapezius, acting to stabilize the fulcrum. By thus rotating the scapula (moving the inferior angle laterally), the trapezius aids the deltoid in its task of abduction of the humerus.

30. Muscles which move the arm: (1) pectoralis major, (2) deltoideus, (3) teres major, (4) latissimus dorsi (the site of insertion described means "at the place where the first three muscles form a continuous line or mass of insertion extending far down the humerus from near its head"), (5) subscapularis, (6) infraspinatus, (7) supraspinatus.

31. The fifth, sixth, and seventh of the group actually reinforce the ligaments of the shoulder joint and are concerned with internal and external rotation.

32. Muscles which move the thorax: (1) subclavius, (2) serratus anterior (modern texts reverse the order of origin and insertion here), (3) serratus posterior superior, (4) probably iliocostales lumborum et dorsi, (5) serratus posterior inferior, (6) transversus thoracis.

33. The enumeration of the intercostal muscles is at first glance somewhat formidable. By the term "true ribs" (costae verae) we mean the seven which connect to the sternum by separate costal cartilages. By "false ribs" (costae spuriae) we indicate the

eighth, ninth, and tenth, all sharing a cartilaginous connection, and the eleventh and twelfth "floating ribs," devoid of cartilaginous segments. There are, then, six inter-spaces for the true ribs and five associated with the false ribs. (The distinction be-tween true and false ribs can be traced back at least as far as Gerhard's translation of Avicenna's *Canon;* see Singer's *Evolution of Anatomy,* p. 81.)

Each of the six interspaces is occupied by an external and an internal intercostal muscle. Vesalius subdivides each layer into two muscles: first, the fibers between the ribs proper; and second, the fibers between the costal cartilages. He thereby obtains four muscles per interspace. This division has validity for the internal intercostals (Wiggers' *Physiology in Health and Disease,* in the chapter on "Respiratory Move-ments and Mechanics of Lung Inflation," differentiates internal intercostals into interchondral and interosseous parts on the basis of functional differences during respira-tion), but the interchondral portions of the external sheet are aponeurotic and non-muscular "external intercostal ligaments" in modern conception (Morris, *op. cit.,* p. 488). Nevertheless, it being clear that Vesalius regards them as muscles, we find four muscles per interspace in six interspaces, a total of twenty-four muscles.

In the five interspaces associated with the false ribs, there are no appreciable inter-chondral spaces, with consequent lack of muscles in this category. Therefore, there exist only the two (external and internal) between each rib, a total of ten in all five interspaces.

The figure thirty-four is the sum of twenty-four plus ten muscles. To list forty, include the six muscles listed in note 32; doubling this forty (to include both sides), we obtain the eighty.

34. The term "septum transversum" is now an embryologic one; the diaphragm is related to it but the two are not identical.

35. The central tendon.

36. The relationship of the alimentary canal to hematopoiesis is set forth in Chap. III; this, plus the liver-spleen system of blood formation and the urogenital system, completes the abdominal contents. The lungs minister to the heart in several ways; according to Plato (one of Vesalius' authorities), they even serve mechanically as a shock absorber (*Timaeus*). We speak frequently of the cardiorespiratory system.

37. Muscles of the abdomen: (1) obliquus externus abdominis, (2) obliquus internus abdominis, (3) rectus abdominis, and (4) transversus abdominis. Observe that in this text no light is thrown on the superior attachment of the rectus abdominis. Vesalius' plates on muscle in the *Fabrica* have been criticized as showing this muscle ascending as high as the manubrium.

38. Compare here the training of singers, speakers, and wind-instrumentalists in dia-phragmatic and abdominal breathing, rather than intercostal. Vesalius seems to attrib-ute to these muscles only the accessory respiratory function, along with a passive containing of the abdominal viscera. Other functions, such as in efforts of expulsion (childbirth, defecation) and in lifting of weights, have been neglected.

39. Muscles which flex the elbow: (1) the anterior, biceps brachii, (2) the posterior, brachialis, probably including coracobrachialis with the short head.

40. Muscles which extend the elbow: the triceps brachialis—(1) long head, (2) lateral head, (3) medial head. The "posterior process" of the ulna is the olecranon.

41. The palmaris longus; see note 5 of this chapter on "true skin." It is worth while to note here the first use of a device employed several times in subsequent text; that is, the contrast of "it is believed" to "the professors of dissection believe." The phrases are so used as to appear that in neither case is Vesalius as yet prepared to make a posi-tive statement; however, in the first case he suggests in an affirmative sense, while the second is indicative of doubt and partial denial.

42. Muscles which pronate the radius: (1) pronator teres, (2) pronator quadratus. The forearm is being described in the anatomical position (radius and ulna parallel, thumb out, and palm anterior).

43. Muscles which supinate the radius: (1) brachioradialis (old name, "supinator longus"), (2) supinator (supinator radii brevis).

44. Muscles which move the brachial (wrist): (1) flexor carpi radialis, (2) flexor carpi ulnaris, (3) extensor carpi radialis longus et brevis, (4) extensor carpi ulnaris. Several points of confusion and ambiguity enter in attempting to define this group.

45. Muscles which move the fingers: (1) flexor digitorum sublimis; (2) flexor digi-torum profundus; (3) flexor pollicis longus (for "third joint of the thumb" see Chap. I, note 45); (4-16, incl.) the text is not adequate for exact identification, but the group probably includes for the thumb the extensor pollicis brevis, abductor pollicis brevis, flexor pollicis brevis (lateral head), opponens pollicis, and adductor pollicis (transverse head); for the index finger, one dorsal and one volar interosseus; and for the little finger, a volar interosseus and probably the flexor digiti quinti brevis and opponens digiti quinti considered together; (17) extensor digitorum communis; (18) extensor digiti quinti proprius; (19) extensor indicis proprius; (20) abductor digiti quinti; (21) extensor pollicis longus et brevis; (23) flexor pollicis brevis (medial head, also sometimes described as interosseus volaris primus); (24) adductor pollicis (oblique head) [?]; (25-28) lumbricales.

   Some of the difficulty in understanding this section arises from inconstant use of the anatomical position in determining direction.

46. The tunica vaginalis.

47. Cremaster muscle.

48. Ductus (vas) deferens.

49. This is probably not the broad ligament and uterine tube. While Vesalius did not recognize the significance of the tubes until they were described by Fallopius, he did consider them as uterine horns, as indicated in his letter to Fallopius. More likely he means here the smooth (involuntary) muscle which may be found in the utero-sacral ligaments.

50. Sphincter urethrae.

51. Sphincter ani; the internus and externus are probably not differentiated.

52. Levator ani.

37

53. The original text uses extreme emphasis in the idea of "very slowly"; the sense is that of "lazily." Vesalius seems uncertain of the effectiveness of the muscle. This is probably the ischiocavernosus muscle (which does have some pubic origin) although the ambiguity might admit identification of a ligament of the penis, such as fundiform or suspensory.

54. Bulbocavernosus muscle. Present ideas of function attribute an ejaculator mechanism to this muscle.

55. Muscles which move the thigh: (1) glutaeus maximus (which passes over the greater trochanter with the interposition of a bursa and inserts instead just below on the gluteal tuberosity), (2) glutaeus medius, (3) glutaeus minimus, (4) piriformis, (5) the adductor group (including brevis, longus, magnus, and probably quadratus femoris—the upward continuation of the adductor magnus), (6) psaos major, (7) iliacus, (8) pectineus and adductor longus, perhaps including brevis, (9) obturator externus (which turns the femur *outward*), (10) obturator internus (which is "increased" by the superior and inferior gemelli).

56. Muscles which move the leg: (1) sartorius, (2) gracilis, (3) semitendinosus, (4) biceps femoris, (5) semimembranosus, (6) tensor fasciae latae (the membrane is, of course, the fascia lata proper), (7) vastus lateralis, (8) vastus intermedius et medialis (except for the lower portion of the medialis, which has a slightly different function, the consideration of these two as a unit is as defensible as the modern separation of them), (9) rectus femoris. The statement that the last-named "turns into" a tendon has the force of "degenerates"; following the Aristotelian idea of more and less noble parts, muscle is superior to tendon, just as bone is to cartilage.

57. Popliteus.

58. Muscles which move the foot: (1 and 2) gastrocnemius (medial and lateral heads), (3) plantaris, (4) soleus, (5) tibialis posterior, (6) tibialis anterior (into the base of metatarsal I), (7) peronaeus longus, (8) peronaeus brevis, (9) peronaeus tertius (quite accurately defined; see later the fourteenth of those which move the toes—extensor digitorum longus, of which this is a part; note 60 below).

59. The insertion of the muscle is into the navicular bone. Probably the "bone resembling a die" is the cuboid, in spite of the fact that the talus is also contiguous and that its old name is astragalus (ἀστράγαλος, "a die"). This identification is based on the fact that the tarsus is indicated and that to Vesalius the talus or astragalus is not a part of the tarsus proper (see Chap. I, note 53). In either case, the "die" is not used in the sense of printing or mechanics but should be compared to another old name for talus, πεσσός, a small stone or other substance for playing at draughts. Note also the old Roman game of "knucklebones."

60. Muscles which move the toes: (1) flexor digitorum brevis (with the plantar aponeurosis); (2) flexor digitorum longus; (3) flexor hallucis longus; (4-13) for the first toe the flexor hallucis brevis and adductor hallucis; second toe, the two interossei dorsales; third toe, one interosseus dorsalis and one interosseus plantaris; fourth toe, the same; fifth toe, plantar interosseus and flexor digiti quinti brevis; (14) extensor digitorum

longus; (15) extensor hallucis longus; (16) extensor digitorum brevis; (17) abductor digiti quinti; (18) abductor hallucis.

61. Quadratus plantae and lumbricales.

62. Capsular ligament and collateral ligaments.

63. The apical odontoid ligament.

64. The transverse ligament of the axis.

65. Intervertebral fibrocartilages.

66. That is, they are not specialized accessions; falling in the class of the first-named they are capsular ligaments, between articular processes.

67. The interspinous ligaments, compared to the interosseous membranes of forearm and leg.

68. The obturator membrane.

69. The coracohumeral is the first; the next two are glenohumeral; and possibly the last of the group is that portion of the glenohumeral which arises from the base of the coracoid process and attaches near the lesser tubercle.

70. Sacrospinous and sacrotuberous ligaments.

71. Ligamentum teres femoris.

72. Menisci and collateral ligaments.

73. The various retinacula and mucous tendon sheaths are not differentiated here, as appears by inspection of the text.

74. Transverse carpal ligament and digital extension of palmar fascia.

75. The six tendons sheathed on the dorsum of the wrist are, from medial to lateral: extensores carpi ulnaris, digiti quinti proprius, digitorum communis and indicis proprius, pollicis longus, carpi radialis longus et brevis, pollicis brevis and abductor pollicis longus.

76. Transverse crural ligament (and mucous sheath of tibialis anterior muscle).

77. Mucous sheaths for tibialis posterior, flexor digitorum longus, and flexor hallucis longus.

78. More likely the mucous sheath of the peronaeus longus than the retinaculum peronaeorum superius.

CONCERNING THE ORGANS WHICH MINISTER

TO NUTRITION BY FOOD AND DRINK

CHAPTER III

SINCE man has been unable to form the substance of an immortal being by means of the genital semen and the menstrual blood (the origins of our generation and of those parts of which we are composed [1]), the great Creator of things has carefully devised that man should live as long as possible and that his species, never failing, should continue to exist forever. In order that man might attain to the stature for which he was intended and that those elements upon which his innate heat is continually fed might be restored as quickly as possible, he possesses organs which serve to nourish him in many ways.

The food is broken up by the teeth in order that the task may later be completed more easily. Food, as well as drink, passes from the mouth to the stomach as into a storehouse along a path called the esophagus or gullet [2]. This is extended by two special tunics appropriately formed to descend from the fauces behind the rough artery [3] and then along the vertebrae of the thorax through the transverse septum to the upper, or left-hand, orifice of the stomach.

The stomach lies between the liver and the spleen under the septum. It is particularly roomy and rather long transversely, larger on the left-hand region of the body than on the right; it is equipped with two tunics suitable for distending and contracting [4] and enclosed by a third covering derived from the peritoneum. The stomach is intertwined with many veins, arteries, and nerves. It concocts what is sent down to it from the mouth and changes this by an innate force into a thick milky juice [5]. This passes through the lower orifice of the stomach from the higher region of its right side and is sent into the intestines.

40

The intestines are rounded, extending from stomach to anus in a continuous course made tortuous by innumerable coils and turns; like the stomach, they are fashioned from two tunics. To these is added from the peritoneum a third tunic adapted for relaxing and contracting no less than the first two tunics proper to the intestines but not everywhere equally extensive [6]. The origin of the intestines proceeds from the stomach; along the posterior side of the stomach, reflected toward the back, lies the organ we call the duodenum. Following this is the part of the intestines known as the jejunum and that which is called the ileum or volvulus. The coils of the latter fill the ilia and the region lying under, and contiguous on all sides to, the umbilicus; it is of almost constant diameter. The narrowness of this organ provides the reason for designating as small intestines the parts just mentioned. The part of the intestines in which the terminus of the ileum lies is broad and very thick; in its entire course it constitutes the colon. Joined to it is a small appendage, narrow and curled like an earthworm; this has one orifice and is therefore called blind by the masters of dissection [7]. The thick part of the intestines itself ascends from the region of the right kidney to the concavity of the liver. Thence it proceeds along the base of the stomach to the region of the spleen, then turns downward along the region of the left kidney, and bends back to the left region of the pubis in a sort of coil [8]. This last passes above the beginning of the os sacrum straight down to the anus, thereby obtaining the name of the straight and principal intestine [9].

Thus, whatever has been prepared in the stomach is sent down through these intestines to be forced through their various coils. Veins in innumerable series pass from the concavity of the liver, together with the arteries drawn off from the great artery, between the two membranes which fasten the intestines to the back. These veins are quite thick and dense, abounding in much fat and glands; they are called the mesentery and extend to the intestines [10]. The veins suck out from the intestines (especially the small ones) whatever is suitable for the making of the blood [11], together with the aqueous and thin refuse of the stomach's concoction, and carry it to the workshop of the liver, where the blood is made. But that refuse which is thicker and less adaptable to suction is gradually collected in the thick intestine; it is kept there only until, it becoming troublesome to man, the muscle surrounding the rectum in circular fashion relaxes, and the refuse is borne forth at once and completely at the will of man.

The liver is not divided into fibers or lobes [12]; it occupies a position higher than that of the organs ministering to it [13] and for the most part lies in intimate relation to the stomach; the liver is placed close beneath the transverse septum and fills the right, rather more than the left, region of the body. It is gibbous above and hollow below, conforming exactly to the shape of the parts lying near it [14]. It is formed by the intertwining of many veins and is surrounded by the substance proper to the liver, similar to recently coagulated blood [15]. It is clothed with a thin membrane [16] proceeding from the ligaments with which it is secured to the peritoneum. It admits two small nerves [17] and one artery. It is the tinder of the natural or nutritive faculty or, as Plato said, of the part of the soul which desires the pleasures of love, food, and drink [18].

One series of veins diffused through the liver lies in its gibbous part, extending to the vena cava [19]. Another series forming the stem of the portal vein lies in the hollow of the liver. This vein sends two branches first to the bladder which receives the yellow bile [20], then to the lower region of the stomach near its lower orifice [21].

Thence a branch runs to the right part of the base of the stomach [22], from which small branches spread out to the stomach and the upper membrane of the omentum [23]. The omentum is a membranous body fashioned like a sack and especially adapted for conducting vessels in safety. However, since it is full of veins, arteries, and the fat attached to them, it also assists in preserving the warmth of the intestines. It is borne in a circular fashion, beginning from the middle of the back under the posterior region of the stomach, through the hollow of the liver, to the base of the stomach (from the third tunic of which it here arises). Then it is carried down to the hollow of the spleen and thence to the middle of the back where it started [24]. Like a sack stretched downward, the omentum covers the anterior region of the intestines, or there where the colon is stretched under the stomach, it arises, joining the back in place of a mesentery. The stem of the portal vein, after having been supported by the omentum, sends out the branches just mentioned.

The stem is divided into two trunks; the right one [25], which is larger, is carried in various ways through the mesentery and is offered to the intestines (first the duodenal intestine); the right trunk also presents a branch to the beginning of the jejunum. The right trunk is supported by a glandulous body stretched out in this region of the intestines [26]. The left trunk, having been woven into the lower region of the omentum, sends a small branch to the posterior region of the stomach, where the latter faces the right part of the back, then also to the inferior membrane of the omentum. Next the branch goes to the glands, fleshy in color, which are in charge of the safe distribution of the vessels here. A branch ascends from it along the posterior side of the stomach; this branch sends out other branches to the region of the stomach which faces toward the middle of the back and embraces the upper orifice of the stomach in the manner of a crown [27]. From this, in addition to the branches sent upward and downward, one creeps forward along the posterior side of the stomach to its lower orifice.

The left trunk of the stem of the portal vein, extending ever to the left [28], sends an outstanding vein woven into the omentum and the colon; it is divided into various branches and sends an offshoot as far as the lower membrane of the omentum [29]. It is inserted into the hollow of the spleen by means of its own offshoots before they enter the spleen. To the left side of the stomach, it sends little branches [30], among which a notable one creeps along the base of the stomach in the left region and sends shoots to the stomach and to the upper membrane of the omentum [31].

Offshoots of the portal vein are distributed through the substance of the liver; within them is contained whatever is brought to the liver from the intestines, to say nothing of the stomach. The liver, concocting the best part of that chyme, changes it into blood, obtaining also a twofold refuse of its concoctions, such as we see in all wines and other similar concoctions. One is thicker than the other and, because it is considered, as it were, the dross and offscouring of the blood, is commonly called the black bile. It is carried through the portal vein to the spleen, which lies below and behind the left side of the stomach. The spleen looks like a rather thick tongue [32]; it adjusts itself to the shape of the organs lying close to it, just as the liver does. It is likewise interwoven with many veins and arteries, by which the proper flesh of the spleen is rendered similar to muddy blood. The spleen is covered with a thin tunic sent forth from the omentum [33]. We believe [34] the spleen draws to itself the thicker refuse of the liver and converts it into nourishment for itself, and whatever it cannot assimilate, it throws up into the stomach.

The thinner refuse of the liver, which is regarded as a sort of flowers of the wine, is the yellow bile. It is drawn into passageways between those offshoots of the portal vein and the vena cava which are distributed through the substance of the liver. These passageways, gathered together, end in a single channel which proceeds from the hollow of the liver and extends to the gall bladder [35]. This, like a rather long pear in the concavity of the liver, arises in the middle of the liver's breadth and is provided with a body adapted for distending and relaxing. The professors of anatomy are convinced [36] that, in the case of this bladder, the bile is preserved until, by the action of its special duct, it is thrust forth into the duodenum. The bile must be carried out along with the dry refuse of the stomach. With its biting quality it irritates the intestines for propelling this refuse and frees them from the phlegm which clings to the refuse [37].

The blood, cleansed of the excrement just referred to, rushes from the narrowest branches of the portal vein into the smallest offshoots of the vena cava [38]. The blood uses as a vehicle [39] the thin watery refuse which it had taken up to the liver from the intestines. This refuse, accompanying the blood thus far and ascending together with it into the vena cava, renders a signal service to it in these narrow passages. For since, thus far, this refuse aids the blood in the function of a prompt distribution, it is also suitable for this refuse to carry off whatever overabundant supply of itself [the refuse] the blood does not require and to purge the blood of that which would be a burden.

This office of purgation is most fitly performed by the kidneys, one each on either side of the vena cava and very close to the liver. They quickly draw the greater part of the serous humor of the liver toward themselves and strain it from the blood [40]. In order that they may accomplish this more handily, a notable vein and likewise an artery are extended to the kidney; the kidney receives the serous blood into a membranous sinus which is broad and hollow and divided into many offshoots concealed by the substance of the kidney and covered over with a double tunic [41]. By its function the urine is expelled and led off into another sinus which is prolonged as the urinary passage constructed like a vein [42]; this urine is going to be carried to the bladder.

The bladder, situated at the posterior region of the pubic bone, gradually receives the urine [43]. Shaped like a rotund flask, it is formed of its own simple and sinewy tunic, interwoven with a threefold type of fibers, ready to be distended or contracted [44]. Another membrane is drawn over it from the peritoneum or the membrane of the abdomen, which is the covering and protection of the organs thus far mentioned. Single passageways from each of the kidneys are carefully inserted in the posterior portion of the bladder not far from its neck. This collects the urine only so long until it troubles man either by its abundance or its quality; then it is completely voided by the opening of the muscle which surrounds the neck of the bladder in circular fashion [45].

The blood, purified by this operation, is distributed through the branches of the vena cava or its rivulets over the entire body in order that the separate parts may drain from the blood that which is proper to them; changing and applying it to themselves, they then convert it to their own nourishment. Finally, also, they drive off the refuse of this concoction from themselves by their own functions.

The series of the vena cava is for the most part as follows: while it is located in the posterior region of the liver, it sends forth branches from its own anterior aspect distributed in a numerous series to the gibbous part of the liver. Then ascending and perforating the transverse septum and the pericardium, it sends two offshoots to the septum [46]. At the level of the right auricle of the heart, the vena cava opens toward the right ventricle with an opening wider than the circular width of the vena cava elsewhere. From the posterior region of its implantation (unless you prefer to say rising [47]), a vein proceeds surrounding the base of the heart in the manner of a crown and sending little branches downward along the upper surface of the heart [48]. The vena cava, rising upward from the heart, there where it pierces the pericardium, sends off from the right side, the azygous

vein, which nourishes the eight, more commonly lower, intervals of the ribs *
on both sides.

In the throat the vena cava is divided into two parts [49], sending veins
from its anterior region to the pectoral bone and to the membranes dividing
the thorax, and creeping through the upper part of the abdomen [50]. From
the root of the one branch of the division into two parts in the throat arises
a notable vein running above the first rib to the armpit [51] but first sending
into the cavity of the thorax a branch which disappears in the three upper
intervals of the ribs of its side [52], and another branch through the transverse processes of the cervical vertebrae all the way to the skull [53], and
another spread out in the posterior part of the thorax near the root of the
neck [54]. The present vein, having meanwhile, passed out of the thorax,
here sends forth the shoulder vein [humerariam uenam] [55] and a branch
to the muscles spread over the anterior region of the thorax [56]; then,
hastening on into the armpit, it sends another to the posterior region of
the thorax and the hollow of the scapula [57], and then another to the side
of the thorax [58].

The remaining branch of the trunk split into two parts in the throat
is again divided into two unequal branches. Of these, the inner and more
slender one forms the internal jugular vein. It enters the skull, with two
offshoots passing to the dural membrane of the brain. The outer branch
sends an offshoot from its outer side; from this the humeral vein [humeralis
uena] is derived [59]. It goes upward, forming the superficial jugular,
running in various ways up to the fauces, and distributed to the tongue,
the larynx, the palate, the face, the temples, and the vertex and entering
the skull with three veins.

The humeral vein, before it is carried under the clavicle and the acromion
into the arm, extends a branch to the posterior region of the neck [60]
and another to the gibbous part of the scapula. Another branch goes to the
upper region of the acromion [61], and a second creeps under the skin,
following along the outer side of the anterior muscle of those which flex
the elbow [62]; bringing forth slender shoots to the skin, it divides in front
of the elbow joint. Sometimes one branch lies deeper and soon disappears,

* . . . . octo frequentius inferiora costarum intervalla.

passing to the elbow joint. Another runs obliquely under the skin to the middle of the bend of the elbow [63] to meet with the branch of the axillary vein and to form one vein in common with it [64]. A third passes under the skin along the radius to the posterior aspect of the forearm, finally to the root of the wrist near the end of the ulna [65]; it mingles there with the offshoot of the axillary vein and rises to the outer side of the little finger and the ring finger [66].

The axillary vein lies hidden in the armpit and sends a branch to the skin covering the anterior region of the arm toward the inner aspect. It presents an offshoot to the heads of the muscles which extend the elbow and another at almost the middle of their length [67]. Then the axillary vein sends another offshoot, with the fourth nerve that proceeds to the arm, along the posterior aspect of the arm up to the exterior region of the forearm [68]. It is soon cut into two veins, one of which sinks completely into the depths in its entire length, continually accompanied by an artery [69]. This vein passes through the middle of the bend of the elbow joint; before it reaches the middle of the length of the forearm, it is cut into two offshoots, one of which stretches along the radius [70], the other along the ulna toward the wrist [71]. Here it is again split into offshoots [72], distributed to the inner region of the fingers in such a way that it offers two twigs to each of them; one shoot, extending to the external region of the hand, is distributed between the first internode of the thumb and the metacarpal bone supporting the index finger.

Extending under the skin all the way, the other branch from the axillary vein [73] is divided into two branches near the elbow joint. One of these branches runs obliquely toward the bend of the elbow joint and merges with the branch of the humeral vein which is composed of those two middle veins, forming a common vein with it [74]. Running obliquely downward along the radius, this is divided into two offshoots like the letter Y in the external region of the forearm. One of these runs for the most part to the external region of the middle finger; the other runs to the thumb and index finger and sends an offshoot into the internal region of the hand to mingle with the small branches encircling the sacred hill of Venus [75]. The other branch from the axillary vein formed by the division near the elbow joint

sends various offshoots to the internal region of the forearm [76]. In associa-
tion with these branches, a vein often occurs arising from the other branch
constituting a common vein which the axillary sends forth. These offshoots
come close together in various ways at times; again, they separate in turn
and are interwoven in the skin in the inner aspect of the forearm. Finally
they creep forward to the skin of the internal part of the hand. The more
outstanding offshoot of this branch extends to the ulna and sends offshoots
in similar fashion into the external region of the forearm. It merges with
the branch of the shoulder vein [humerariae] near the root of the wrist as
that branch runs to the little finger and the ring finger.

A part of the vena cava runs downward below the liver. It sends a branch
from the left side to the fatty tunic of the left kidney and the region contiguous to it [77]. Then a large vein is borne to each of the kidneys. From
the superior aspect of the vein seeking the right kidney (which vessel frequently originates higher than the vein belonging to the left kidney), an
offshoot approaches the fatty tunic of the right kidney [78]. From the
inferior aspect of that vein which passes to the left kidney, a seminal vein
arises; the right seminal vein originates much lower down from the trunk
of the vena cava [79]. Further, where the vena cava lies upon the lumbar
vertebrae, it gives offshoots to the latter in clusters which finally disappear
into the nearby muscles and sides of the abdomen [80]. The most outstanding of these are those which arise from the vena cava where it divides
into two equal trunks [81] above the union of the os sacrum and the
lumbar vertebrae. Both the right and left trunks send some offshoots to the
foramina of the os sacrum [82].

Each trunk is divided into two branches, of which the inner [83] sends
an offshoot which ends in the muscles occupying the posterior regions of
the iliac and sacral bones [84]. Another offshoot goes to the bladder and
penis [85]; in women it extends to the uterus in the form of many smaller
shoots [86]. That which is left of this branch anastomoses with the external
branch and is led through the foramen of the pubic bone to the thigh [87],
where it sends offshoots to the skin and muscles occupying the inner femoral
region. Proximal to the knee joint, it ends, joining its terminus with the
branch of another vein which extends to the leg, as I shall soon describe.

The external branch [88] of the left trunk of the vena cava, when it is about to pass through the groin to the thigh, sends to the peritoneum an offshoot which terminates in the lower region of the abdomen up to the umbilicus [89]. Extending downward upon the thigh, the external branch sends a shoot to the skin of the pubis and to the hillocks of the female pudenda [90]. It sends a notable vein under the skin through the internal aspect of the thigh, knee, and leg as far as the end of the toes [91]; in its progress it distributes other branches here and there to the skin. Another vein is also sent under the skin to the anterior region of the hip joint [92]. Itself more deeply submerged among the muscles, the trunk [93] sends an offshoot to the muscles located in the external region of the thigh and to the skin [94]; it sends another offshoot to the muscles which appropriate to themselves the inner and anterior region of the thigh [95]. With this offshoot is joined the end of that vein which descended through the foramen of the pubic bone. Thence the large vein [93] winds back to the posterior part of the thigh and sends offshoots to the muscles of that region; from these offshoots little branches extend to the skin, upward and downward as far as the calf.

This large vein is divided into two trunks between the lower heads of the femur. The lesser outer one extends to the fibula [96]; from it, in addition to small branches extending to the anterior aspect of the knee [97], a branch is separated which proceeds posteriorly under the skin covering the external region of the leg and which is variously divided toward the upper part of the toes [98]. That portion of it which remains hidden higher up among the muscles extending toward the external region of the fibula runs past the middle of the length of the leg. The inner of the two trunks is quite large [99]; along the inner region of the tibia, it sends a branch spread out posteriorly under the skin as far as the toes. Another branch is sent forth, somewhat hidden through the calf and stretching as far as the heel. Especially worthy of note concerning this trunk is the fact that it extends to the muscles which occupy the posterior aspect of the leg. It sends a branch from its anterior aspect down through the membranous ligament which binds the fibula to the tibia. The branch hidden under the anterior muscles which surround the tibia extends to the upper part of the foot [100].

The vein itself, running down along the posterior part and thence sending shoots to the skin and the contiguous muscles, finally enters the lower portion of the foot between heel and tibia and is there distributed to the muscles and toes in such a way that two offshoots are sent to each toe.

GENERAL INTERPRETATION

IN THIS and the following chapters, the physiology departs sufficiently from modern concepts and is so dispersed through the text that it seems appropriate to bring it together in brief abstracts. To these are joined such remarks of general nature as may refer to the subject and yet find no specific place in the serial notes.

The Vesalian discussion of blood and bile is an effort to bring to the prevailing humoral theory of physiology and pathology some degree of anatomical specificity. The theory, which probably originated with primitive man, was used in Vedic medicine, proclaimed to the West by Hippocrates, "proved" by Galen, and held in good repute until the time of Virchow. It is not surprising to find that, in the main, Vesalius accepts it; his position is to carry forward the efforts of Erasistratus to take it out of the realm of metaphysical speculation.

Vesalius believes that the mesenteric vessels carry material absorbed from the intestines to the liver (via the portal vein). Blood is formed in the liver and enters the circulation through the hepatic vein. In the liver is obtained a refuse which consists of two parts, comparable to the flowers (or supernatant scum) and lees (or dregs) of wine. The thicker is the black bile, which is sent back through the portal vein to the spleen; no effort is made to explain this two-way flow in the portal vein (intestine to liver and liver to spleen). It is not difficult to understand the "black bile" (atra bilis, melancholia) of the splenic vein when one remembers that the blood coming from the spleen actually contains a high concentration of hemoglobin breakdown pigments. Anything in the black bile not usable by the spleen is sent (via the left gastroepiploic vein?) to the stomach.

The thinner decoction of the liver is the yellow bile, which is sent to the gall bladder. While Vesalius has some doubt as to the role of the gall bladder as a storage organ, he recognizes that bile is the "physiological laxative," acting as an irritant (through its "biting quality") to increase intestinal motility.

The noncellular component of blood leaving the liver contains some waste material, which is carried to the inferior vena cava and then to the kidneys for excretion. In a later chapter Vesalius states that the two kidneys have slightly different functions with respect to this excretion; he thereby accounts for the difference in right and left internal spermatic veins and explains the method by which erotic activity and formation of seminal fluid are stimulated by its acrid quality.

The blood is now freed from all useless or harmful substances and is distributed to all parts of the body by the veins. The various tissues extract from blood useful substances and return to it their own refuse from metabolism.

Phlegm, the fourth humor, is mentioned only in passing; the account of its origin is contained in a later chapter.

The only authority named in the *Epitome,** Plato, is cited in this chapter in connection with the discussion of the liver. While the exact source is not given, I believe it to be the *Timaeus.* This dialogue, while starting with cosmogony, mathematics, and

* With the exception of a passing mention of Galen in the section on regional anatomy.

astronomy, devotes its last third to anatomy, physiology, and pathology. Without doubt Vesalius is as familiar with the content as he is with the work of Hippocrates, Aristotle, and Galen. Plato believes the soul is tripartite, with special regions of the body as the seat of each part. The head, being spherical and most perfect in shape, is the seat of reason and intellect, the most perfect part of the soul. It is above the other seats of soul-portions, and hence reason is placed in a dominant position. The breast is the seat of the nobler passions; it is not to be identified with the head, and hence the neck has been interposed. The diaphragm divides the nobler soul from the coarser base soul, with its bodily appetites and sensuality, seated in the abdomen. The liver has a degree of control over this last in that: (*a*) it contains "bitter," which it uses to restrain the cravings; (*b*) it contains "sweet," to be discharged when desires conform to reason; (*c*) it is like a mirror, smooth and bright (note the pathologist's phrase, "smooth, moist, and glistening"), and may thus reflect the thoughts.

In the *Symposium* of Plato, the myth of the Charioteer and the two horses is thought to be an allegorical statement of this theory of the soul.

The reader interested in the history of medical concepts will, if not already familiar with the *Timaeus,* find much delight in reading it.

1. BY Aristotelian teaching (as in *De Generatione Animalium*), the semen of the male is a pure secretion containing the soul principle, while the catamenia is a female semen lacking this principle. Introduction of the soul principle to the secretion of the female results in conception and formation of the embryo.

2. The "path" itself is the lumen, and the two tunics the mucosa and the muscularis-adventitia.

3. Trachea; "rough" because of its cartilage rings, "artery" because it is an air passage.

4. The same two layers as in note 2, plus the serosa, a reflection of the peritoneum.

5. The acid chyme.

6. This is a brief statement of the fact that not all of the intestine is in a broad mesentery but in a number of places becomes almost retroperitoneal.

7. The cecum, "blind," as in a "blind alley" or cul-de-sac. The appendix vermiformis is the part here designated in particular as blind.

8. The sigmoid colon.

9. Intestinum rectum.

10. The mesenteric vessels, within the two sheets of serosa making up the mesentery. The glands are mesenteric lymph nodes.

11. "Suck out": compare the observation of H. S. Wells, *Am. J. Physiol.*, XCIX (1931), 209, that the osmotic pressure here is enough to balance a negative intra-intestinal pressure of 8 to 26 cm of normal saline solution. While we would not say "suck," the concept of some of the absorptive mechanism is suggested.

12. In correction of a standing anatomical error of the time, probably deriving from dissection of lower mammals.

13. "Ministering to it" via the portal venous system.

14. The many impressions on the surface of the liver.

15. The Galenic concept of the nature of liver parenchyma.

16. The tunica serosa—Glisson's capsule.

17. From the left vagus and the sympathetic (celiac plexus) via the hepatic plexus.

18. The *Timaeus;* see general interpretation to this chapter; also *Republic* 439a-440.

19. The intralobular veins join to form the hepatic vein.

20. Gall bladder; cystic vein.

21. Pyloric vein.

22. Right gastroepiploic vein.

23. "The upper membrane" is the anterior portion of the gastrocolic fold.

24. Successively, the gastrohepatic, gastrolienal, and lienorenal "ligaments" or portions of the omentum.

25. The superior mesenteric vein.

26. Pancreaticoduodenal vein; the pancreas.

27. The coronary vein.

28. The lienal (splenic) vein.

29. The inferior mesenteric and left colic veins.

30. The short gastric veins.

31. The left gastroepiploic; the vein is "notable" because of its part in the relationship of spleen to stomach (see text and general interpretation).

32. That is, it has a broad side and an opposing sharp border.

33. The capsule is formed by a reflection of the dorsal mesogastrium.

34. See Chap. II, note 41.

35. The bile ducts and hepatic duct.

36. See Chap. II, note 41. To bring the sentence on the gall bladder up to date, one would have only to say that "the professors of anatomy are convinced but the professors of physiology doubt, etc." The words which have been exchanged over just such controversies as this are those which prompted Truthful James to remark:

> I hold it not quite wise in any scientific gent
> To say another is an ass, at least to all intent.
> —BRET HARTE, "The Society upon the Stanislaus"

37. Note the recognition of the laxative function of the bile salts.

38. The confluence of the hepatic arterial and portal venous blood into the central veins and thence via the hepatic vein to the vena cava.

39. Vehicle, in the (present) pharmaceutical sense of a fluid "carrier" for the active constituents.

40. The liver effluent blood contains wastes for excretion by the kidney; for example, the liver is a prime source of urea, one of the discards of protein metabolism.

41. The outer "tunic" was the peripelvic tissue; the inner, the kidney pelvis and calyces.

42. Ureter.

43. Observe the contrast in Vesalius' expressions of rate of activity: blood "rushes" "quickly" in speaking of liver and kidney circulation in the two preceding paragraphs, but urine is secreted "gradually." Vesalius performed a number of physiologic observations as well as anatomies. (Lambert, "The Physiology of Vesalius," *Bull. New York Acad. Med.*, XII, No. 6 [June, 1936].)

44. Nerve: a connective tissue sheet like an aponeurosis; see Chap. II, note 1.

45. Chap. II, note 50. The detrusor muscle and the sphincters.

46. The inferior phrenic veins.

47. Using the Aristotelian idea of a vena caval origin at the heart, rather than the Galenic concept of a hepatic origin.

48. Great cardiac vein and especially the coronary sinus.

49. The innominate veins.

50. Internal mammary and superior phrenic veins.

51. Axillary vein.

52. Highest intercostal vein.

53. Vertebral vein.

54. Superficial cervical vein. From this point in the description of the veins of the upper extremity, it has not proved possible to identify every vessel with certainty. This is due partly to the considerable variability of even larger veins in this region and partly to obscurities in description and form of outline. In consequence, only the vessels which seem most certainly identified are noted.

55. Cephalic vein.

56. Anterior pectoral vein.

57. Subscapular vein.

58. Lateral thoracic vein.

59. Subclavian, axillary, and brachial (in succession).

60. Transverse cervical vein.

61. Acromial and deltoid veins.

62. Basilic vein.

63. Median cubital vein.

64. Median antibrachial vein.

65. Accessory cephalic vein.

66. The most medial of the dorsal metacarpal veins, from the dorsal venous network.

67. Humeral circumflex and muscular rami.

68. Deep brachial vein, with the radial nerve (Chap. V, note 71).

69. Ulnar vein.

70. Dorsal interosseous vein.

71. Volar interosseous vein.

72. Superficial volar arch.

73. Radial vein.

74. Anastomosis with the superficial circulation, especially the median antibrachial vein.

75. The deep volar, with an offshoot to the thenar eminence.

76. The venous anastomosis around the elbow joint, and the cutaneous and muscular rami.

77. Left suprarenal vein; the subsequent account differs from the modern in two main points: (*a*) Modern texts show the right suprarenal as a branch of the inferior vena cava; Vesalius just reverses this. (*b*) Modern texts place the left renal higher than the right; again this is reversed in the *Epitome*.

78. Right suprarenal vein.

79. The internal spermatic veins.

80. Lumbar veins.

81. The common iliac veins.

82. Lateral sacral veins.

83. Hypogastric vein.

84. Gluteal veins.

85. Internal pudendal and vesical veins.

86. Uterine veins and venous plexuses of pelvic viscera.

87. Obturator vein; Vesalius must have followed a branch to an anastomosis with some of the superficial veins of the thigh, perhaps an accessory saphenous.

88. External iliac vein.

89. Deep epigastric vein.

90. Superficial external pudendal vein.

91. Great saphenous vein.

92. Superficial circumflex iliac vein.

93. Femoral vein.

94. Lateral circumflex femoral vein.

95. Deep femoral vein, especially the medial femoral circumflex branch as the anastomotic channel.

96. Peroneal vein.

97. Genicular veins.

98. Small saphenous vein.

99. Posterior tibial vein.

100. Anterior tibial vein.

CONCERNING THE HEART AND THE ORGANS

WHICH MINISTER TO ITS FUNCTION

CHAPTER IV

OF THE organs which are created for rekindling the natural heat within us and for the restoration and nourishment of our spirits, the heart is considered by far the most important part of the agitative faculty. It is like a pine nut, compressed in front and behind, located with its base under the middle of the breastbone and its apex inclining sharply forward to the left side. The substance of the heart is fleshy, but like the substance of muscles, and is interwoven with a tougher, threefold type of fibers [1] provided with its own veins and arteries.

The heart has two sinuses or ventricles [2]. One is located on the right side; this is broader and appears to be covered with a thinner and looser substance of the heart [3]. The orifice of the vena cava extends to this ventricle and is furnished with three membranes drawn inward [4]. Likewise, a vessel which is like an artery in form but performs the function of a vein [5] and hence is called the arterial vein proceeds from this ventricle; this vessel sends toward the orifice of the ventricle also three small membranes facing outward [6]. The other ventricle, surrounded by a special thick substance of the heart, lies on the left side. It, too, has two orifices, of which the lower, with two membranes closing inward [7], belongs to a certain vessel, an artery. While it is formed like a vein, this vessel holds the air and performs the function of an artery [8]; hence it is called the venous artery; this artery sends two membranes that close inward to its own orifice. The higher orifice is dedicated to the beginning of the great artery, to which Nature has also given three membranes facing outward [9]. These ventricles are separated by a very thick septum adapted for distending and contracting and (like the ventricles of the heart) built up within of many pits of ample size [10].

The entire heart is covered with a certain membranous involucrum, to which it is joined at no point [11]. This involucrum is much more ample than the heart and is moistened within by an aqueous humor. The lower region of the involucrum is attached on the outside to a transverse septum of no small breadth; on its two sides the involucrum is contained by the membranes interposed in the cavity of the thorax, supporting this involucrum in order that the heart may be supported in position [12].

The lung fills the rest of the cavity of the thorax not occupied by the heart, the membranes just mentioned, and the descending esophagus. The lung adapts itself on all sides, as the liver does, to the shape of the parts lying close by; on both right and left sides it resembles the hoof of a cow or some other cloven-hoofed animal. Each lung is divided into two fibers or lobes [13] built up from many interweavings of vessels. The rough artery [14] is led down from the top of the throat (where also the tonsils and two other types of glands are located) to the thorax; it is partly cartilaginous so that the voice may be produced. In order that the lung may be expanded and relaxed and thus may assist breathing, it is partly membranous, filled here and there with branches. The arterial vein, proceeding from the right ventricle of the heart which prepares the blood familiar to the lung, offers the blood to the lung; it is distributed in an innumerable series to the latter. Similarly, the venous artery intertwines the lung with an abundant series. These vessels are surrounded by the spongy, soft, foamy, and quite pliable substance proper to the lung. A quite small thin tunic lies next to this substance, not hindering the dilatation and compression of the lung in any way; this tunic is always contiguous to the tunic which lines the ribs [15].

The lung causes a motion of the thorax dependent upon our wills; it dilates to produce a vacuum, and by virtue of this the air from outside ourselves passes along the uvula. When we breathe deeply, the air is attracted through the mouth as though into a bellows. A small part of the air seeks the brain through the foramina of the skull [16], while the remainder [17] enters the rough artery [trachea] by way of the upper throat and completely fills the cavity of the lung made by the latter's dilatation. The substance of the lung changes this air by force peculiar to itself, adapting the air to the use of the heart. This allows the best part of the air to be taken up by the branches of the venous artery from the offshoots of the rough artery extending throughout so that, by the intervention of the former artery, the air may be carried to the left sinus of the heart, where it is going to perfect the material of the vital spirit.*

The heart attracts this air and draws a large supply of blood from the right ventricle into the left ventricle. From the steamy vapor of that blood and from that air, by the inborn virtue of its own substance, the heart creates the spirit which the blood with a rushing flow distributes, thus accompanied and nourished, to the entire body through the great artery; the heart tempers the native heat of each part in the same way that the respiration restores the tinder of the innate heat to the heart. Thus the respiration and the pulse have the same use; by their rhythms the great artery of the heart is dilated and constricted. The heart therefore uses the air for making the vital spirit, and the fiery heat of the heart is tempered by the air. Whatever is sooty and unsuitable for production of the spirit is returned to the lung through the venous artery and, together with the air which had remained in the lung, is driven forth by the compression of the thorax; this is agreed by professors of dissection. To be sure, as the tireless heart by its own dilatation draws the blood into its right ventricle from the vena cava and part of the blood passes to the left ventricle, part of it in fact is appropriately prepared by the heart itself as suitable nourishment for the lung and is offered to the lung through the arterial vein by the contraction of the heart. The dilated heart takes air from the lung into the left ventricle, but when constricted it propels the vital spirit into the great artery with the rushing flow of the blood. In order that the rapid contraction of the heart may not bring harm to the vena cava and the venous artery, Nature has created the auricles as storerooms placed close to the heart [18].

We believe that four membranes guard the orifices of the cardiac vessels so that the heart's labor may not be in vain [19]. The membranes guarding the orifices of the vena cava and of the venous artery prevent the blood from flowing back into the vena cava during the contraction of the heart and prevent the vital spirit from flowing back into the venous artery. Those membranes which guard the orifices of the arterial vein and of the great

* . . . . *spiritus vitalis idonea futurus materia.*

artery prevent the blood carried to the lung and the vital spirit already sent forth from being regurgitated into the heart during its dilatation.

The great artery grows forth from the heart, which it resembles. It sends forth two shoots which girdle the base of the heart and send branches downward through its substance [20]. The stem of the artery divides into two trunks a little above the heart [21]; the larger turns leftward to the spine, and branches [22] extend from it to the eight lower ribs on either side. As this trunk is borne downward below the septum, it sends offshoots to the latter [23]. From one root it sends shoots to the omentum, stomach, liver, gall bladder, colon, and finally to the spleen, it being accompanied by the branches of the portal vein [24]. Then another trunk here sends a root to the mesentery [25] and one to each kidney somewhat lower [26]; seminal arteries are sent forth from the anterior region of the trunk [27]. Then another branch lower down is sent to the mesentery [28]. The trunk, sending offshoots in its course to the lumbar vertebrae and the muscles lying on them [29], arrives at the beginning of the os sacrum; the artery is set on the left side of the vena cava and thus creeps on more safely. It is divided into two parts in the same manner as the vena cava and makes an equal distribution with that vein to the very end of the foot [30]. However, no branch of this trunk of the great artery passes to the skin [31]. That part of this trunk which proceeds separately to the offshoot passing through the foramen of the pubic bone and joins the artery to it, descends from the umbilicus along the side of the bladder; this is regarded as belonging to the fetus [32].

The other trunk of the stem of the great artery goes upward [33]; it soon sends from its left side a branch extending obliquely to the highest rib on this side [34]. From this branch there first goes an offshoot to the upper ribs [35], and next another offshoot to the transverse processes of the cervical vertebrae [36], which finally disappears in the dural membrane of the brain; still another offshoot passes along the left side of the breastbone, is always placed deeply, and extends to the umbilicus [37]. Where the main branch passes over the cavity of the thorax, it sends an offshoot to the muscles which occupy the posterior region of the neck [38]. For the rest, just as the axillary vein is distributed to the very ends of the fingers (if you will except those branches of the vein which reach the skin), so may you understand the arterial branches, which lie only deeply covered here [39].

A sizable portion of the trunk being described ascends to the root of the neck and is divided into two unequal branches [40]. The left and more slender one constitutes the carotid artery of the left side. The right one sends an offshoot from its right side to the first rib [41], where the offshoot completely disappears in the same way in which that former branch of the trunk was said to reach obliquely to the first rib of the left side. The remainder of the right branch forms the carotid artery of this side; like the one on the left, it seeks the upper part of the throat along the side of the rough artery. Deeply into the face it sends a branch which becomes absorbed in the skin of the temples up to the vertex [42]. The artery itself sends offshoots to the larynx, tongue, and the threefold types of glands lying here [43]; it then enters the skull* and, divided into two offshoots, sends the lesser to disappear in the first or right sinus of the dural membrane [44]. The larger offshoot, without an accompanying vein, disappears into the skull through a foramen of its own, and here offshoots soon separate from it and go to the side of the dura [45]. Another offshoot hastens through a peculiar perforation to the cavity of the nostrils toward the end of the nose [46]. But the offshoot itself, not spread out over the base of the skull and not distributed into any intertwining network, passes forward and sends a branch to the eye with the second pair of nerves of the brain [47]; then it ascends, perforating the dural membrane. Here, part of it disappears into the thin membrane [48]; part creeps into the right ventricle of the cerebrum and forms a plexus laid down in this ventricle and compared to the outermost covering of the fetus [49]. The offshoot brings the vital spirit to the brain so that, as I shall now say, the animal spirit may be prepared by the function of the brain.

* Cf. *Fabrica* III, xiv, Tertia Arteriae Series (marginal rubric).

61

THIS chapter deals with the structure and function of the cardiorespiratory system and with the distribution of the arteries.

The heart is described as a two-chambered organ; the atria are not distinguished from the great veins which enter them. The right and left ventricles are the primary chambers of the heart, into which the great veins open. The atrioventricular valves (tricuspid on the right, bicuspid on the left) are placed between the vena cava and the pulmonary veins on the one hand, and the right and left ventricles (respectively) on the other. The auricles (auricular appendages) are chambers close to the heart, serving the function of taking up the shock of systole and thereby protecting the great veins. The right and left ventricles have separate outlets, which are the pulmonary artery (arterial vein) and aorta, respectively. The intrinsic structure of the heart is described as having a muscular wall, thicker on the left than the right, and a septum which shows pits (which, in the *Fabrica,* are denied to be septal perforations or foramina of any degree of importance). The heart is surrounded by a pericardial sac in which fluid is found. The lung is a spongy, bilobate structure which assumes the shape of the space it occupies. It is covered with a pleural membrane, and its substance is made up of the complex intertwining of many branches of the pulmonary artery, pulmonary veins, and bronchi.

The slight hesitance with which Vesalius approaches the question of the preparation and circulation of blood reflects the beginning of the dissatisfaction with Aristotelian and Galenic physiology which was appearing in the scientific world. At the time of this publication, Vesalius' ill-fated contemporary and former fellow student, Michael Servetus, was setting down his ideas on the circulation through the lungs. The crucial points which Vesalius handles diffidently were, perhaps, already taking form in Servetus' mind during the year which the two spent in Guenther's laboratory in Paris (1532), and the manuscript of *Christianismi Restitutio,* in which the passage concerning pulmonary circulation appeared, was written before 1546. At about the same time, Canano of Ferrara informed Vesalius of the discovery of valves in the veins; their significance was missed by Vesalius and Fallopius; the latter probably observed them and communicated the information to his student, Hieronymus Fabricius. (Cf. Fabricius, *De Venarum ostiolis;* translated by K. J. Franklin, D.M. Springfield, Ill.: Charles C. Thomas, Publisher, 1933.) In 1598 the English student, William Harvey, entered Padua and came under the influence of Fabricius; through such a chain of groping uncertainty, teacher may have passed to student the seed of the idea which flowered in 1616 as a report on the circulation of blood—"the greatest discovery in the annals of medicine" (Fulton).

The cardiorespiratory physiology of Vesalius is dependent on the idea of the three-fold "spirit" (Gr., *pneuma*) in man, associated with the three parts of the soul (see Chap. III, general interpretation). The *pneuma physicon,* or natural spirit, is that associated with the lowest or vegetative side of man: it is a vapor formed in the liver and carried with the humors along the veins. The *pneuma zoticon,* or vital spirit, is produced in the heart, the seat of the passions. The source and character of the *pneuma psychicon,* animal spirit, will be examined in connection with Chapter V.

According to Vesalius, the lung expands by activity of the thorax comparable to a bellows, and air rushes in. While some inspired air passes to the cerebrum by way of perforations in the skull, much of it goes through the trachea to the lung. The lung acts on this air and a part of it passes by way of the pulmonary vein to the left ventricle, where the vital spirit is prepared through the interaction of the air and the natural-spirit–bearing blood from the liver; the nonusable part of the air drawn by diastole to the left ventricle returns to the lung during systole by the same route it came and is expelled by thoracic compression in expiration. The blood, now bearing vital spirit, also leaves the left ventricle during systole and passes through the aorta; that which reaches the brain is transformed to animal spirit.

It is in getting blood to the left ventricle that the discussion encounters the greatest difficulty; it is so great, in fact, that Vesalius omits any attempt at explanation in the *Epitome.* The only account given here of the movement of blood which has entered the right heart states that a part of it is sent, in systole, through the pulmonary artery for the nourishment of the lung substance. An explanation is found in the *Fabrica,* where we are informed that air is drawn from the lungs through the pulmonary vein by the left ventricle, whence it "together with the blood which soaks plentifully through the septum from the right ventricle into the left may be assigned to the great artery (the aorta) and so to the whole body." Later, in denying the existence of perforations in the interventricular septum, he says, "We are driven to wonder at the handiwork of the Almighty, by means of which the blood sweats from the right into the left ventricle through passages which escape human vision."

Several excellent references relating to the development of ideas on the relationship between heart and lungs may be found in Sir M. Foster, *Lectures on the History of Physiology* (1901), from whose first lecture the extracts from the *Fabrica* are taken; in J. F. Fulton's *Selected Readings in the History of Physiology* (1930), especially Chapter II; and in Arturo Castiglioni's *History of Medicine* (1941), where the *pneuma* is described (p. 221). See also Samuel W. Lambert, "A Reading from Vesalius and the Physiology of Vesalius," *Bull. New York Academy Med.* (1936), vol. 12, pp. 345–415.

NOTES

1. UNDERSTANDING of what Vesalius meant by the "threefold type" mentioned in Chapters II and III and twice in the present chapter must await annotation of the corresponding sections of the *Fabrica*.

2. The atria are not considered as parts of the heart; the right atrium was, to Vesalius, a part of the vena cava with "an opening wider than the circular width of the vena cava elsewhere" (see Chap. III, p. 45 and note 47); consistent with this, the left atrium is an expansion of the pulmonary vein. The auricles correspond to the auricular appendages of modern terminology.

3. The wall of the right ventricle is the thinner.

4. Tricuspid valve: "drawn" inward means "directed."

5. That is, carries venous blood (to the lungs); pulmonary artery. See the Leake translation of Harvey's *De Motu Cordis*, footnote 7 to the introductory section.

6. Pulmonary valve.

7. Mitral (bicuspid) valve.

8. That is, carries arterial blood; pulmonary vein. Actually the left atrium; see note 2 above.

9. Aorta and aortic valve.

10. The spaces between the trabeculae carneae.

11. Parietal pericardium.

12. The transverse septum is the diaphragm; the lateral membranes probably indicate mediastinal pleura.

13. The three lobes of the right lung are not recognized.

14. Trachea: see Chap. I, note 21.

15. In order, the visceral and parietal pleurae.

16. For the fate of the "small part of the air" see Chap. V, p. 69.

17. For this and following, see general interpretation to this chapter.

18. The auricles (indicating the auricular appendages) are thought to form a kind of recoil mechanism or shock absorber.

19. That is, to prevent regurgitation and retrograde blood flow.

20. Right and left coronary arteries.

21. The innominate artery and (larger) arch of the aorta. See note 33 below.

22. Intercostal arteries (usually nine pairs).

23. Superior and inferior phrenic arteries.

24. The celiac artery, with left gastric, hepatic, and lienal (splenic) arteries and their branches.

25. Superior mesenteric artery.

26. Renal arteries.

27. Internal spermatic (or ovarian) arteries.

28. Inferior mesenteric artery.

29. Lumbar arteries.

30. Compare this sentence with statements in modern texts (such as Morris, p. 729) that "veins accompany the arteries, and have practically the same relations as those vessels." This, together with the brevity of the present chapter compared with the foregoing chapter on veins, affords excellent illustration of the reversal of emphasis on blood vessels which resulted from the discovery of the circulation of blood.

31. Except for the subcutaneous venous system, the arteries of the extremities correspond to the veins; the superficial veins have no arteriae comitantes.

32. The hypogastric artery, which has the obturator as a branch, is much larger in the fetus, and at birth only its pelvic portion is left; the remainder, "which descends from the umbilicus," is the obliterated hypogastric artery, or lateral umbilical ligament (a fibrous cord).

33. The following text appears confused at first; examination of a Vesalian plate of the arterial system (such as may be found preceding p. 65 in *Icones Anatomicae, Tabulae Libri Tertii*) will establish the point of view from which the description pro-

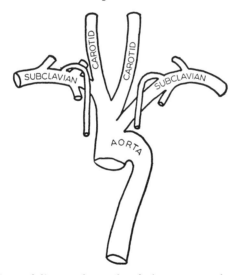

ceeds. Essentially it is as follows: the arch of the aorta and its descending portions have been described as the larger of the two trunks of the stem of the great artery which comes from the heart. The stem is considered to include the ascending aorta and that portion of the arch which gives rise to the great vessels of the head and upper extremities; the aorta, as already described, starts at the point where the left sub-clavian is sent forth. Consequently the lesser of the two trunks, which is now to be described, includes not only the innominate artery but also the left common carotid and subclavian arteries and that portion of the aortic arch from which they arise. The adjacent diagram, adapted from the plate referred to, may aid here. If this concept of the aorta and its branches requires any defense, Morris (*loc. cit.*, p. 607), may be cited: "The innominate and left carotid arise close together—indeed, so close that, when seen from the interior of the aorta, the left subclavian arises a short distance beyond the left carotid."

34. Left subclavian artery.

35. Superior intercostal artery.

36. Vertebral artery.

37. Internal mammary artery.

38. Deep cervical artery.

39. See note 31 above.

40. The two common carotids; there is no named equivalent to the "sizable portion of the trunk" which includes the innominate artery and a portion of the aortic arch.

41. Right subclavian artery.

42. Not only the superficial temporal but the major part of the external carotid artery.

43. Superior thyroid and its laryngeal rami, lingual, and internal maxillary arteries. While positive identification of the "threefold types of glands" cannot be made from the *Epitome,* inspection of *Fabrica* VI, v, suggests that the first type is the thyroid gland, the second is the palatine tonsils, and the third a group of glands which includes the cervical lymph glands and all of the major salivary glands. There is also a refer-ence to what may be the thymus.

44. From this point the description is confusing and inadequate. The present vessel may be the anterior meningeal artery.

45. Middle meningeal artery.

46. Ethmoidal arteries(?).

47. Ophthalmic artery.

48. The cerebral arteries (anterior and middle). The arterial circle at the base of the brain is not described.

49. Choroidal artery and choroid plexus (compared to the chorion).

CONCERNING THE BRAIN AND THE ORGANS

FORMED FOR FURTHERING ITS FUNCTIONS

CHAPTER V

THE brain, the seat of the animal and the principal faculty, lies in the skull and admirably fits the form of the cavity in the upper region of the head which it occupies throughout its length. Anteroposteriorly it is divided into right and left parts but is continuous at its base in the mid-line. Here is the beginning of the dorsal medulla [*oblongata et spinalis*], entirely different from the medulla [marrow] of the bones. Rather less than a tenth of the brain [1] and completely beneath it in its posterior aspect, the cerebellum is joined to the dorsal medulla. The cerebellum extends backward no farther than does the brain itself.

A tough membrane [2] surrounds all these parts of the brain, closely lining the skull and sending fibers through its sutures which end in the skull's own involucrum [3]. This membrane is separated only far enough from the thin membrane of the brain that it may not hinder the motions of the latter's vessels [4]. It sends a process between the right and left parts of the brain [5] and likewise another between the superior region of the cerebellum and that aspect of the cerebrum which rests upon the cerebellum [6]. Performing the service of veins and arteries at the same time and distributing a varied series of vessels to the thin membrane of the brain, four special sinuses lie in this membrane [7]. The substance of the brain, which is continuous, white, and intertwined by no veins, is covered over very intimately by this thin membrane [8]. Weaving here and there through the convolutions of the cerebrum, which are very much like the windings of the intestines [9], it contains the cerebral vessels.

The cerebrum is furnished with three outstanding and spacious cavities or ventricles; of these, one is located in the length of the right part of the cerebrum. It bends back from the posterior portion through the substance of the cerebrum and is carried to the middle of the base of the brain [10]. The second is located in the left part of the cerebrum in a similar fashion. Where they face each other inwardly, they are separated in the upper region by a certain thin substance of the cerebrum [11]; this is called the septum and is continued superiorly to that part of the cerebrum which, because it is somewhat harder and more whitish than the other parts of the brain lying on the surface, is called the corpus callosum [12]. The lower region of the septum is joined to, and continuous with, that part of the cerebrum which is formed like an arch or a tortoise shell (or vault) [13]. It proceeds with a broad base [14] on both sides from the posterior region of the two first ventricles of the cerebrum; gradually passing forward, it fuses at an acute angle and in its inferior aspect overhangs the cavity to be discussed, like the hollow of an arch.

The lower regions of the ventricles mentioned are not partitioned from each other by the septum but come together in a common sinus [15] lying under the body formed like an arch; this space extends directly down-ward as a prominent channel through the substance of the cerebrum into a funnel or basin formed by a thin membrane of that shape [16]. The phlegm of the brain descends through this channel; it is distilled by a quadrate gland lying on the bone shaped like a wedge [17] and thence flows down to the palate and the broad part of the nostrils through notable perforations, which, however, are not like the holes of a sponge [18]. This present cavity, common to both right and left ventricles, is the third ventricle of the brain. It ends posteriorly in a canal [19] which reaches through the bodies of the brain which are not unlike the nates and the testes [20] to the fourth ven-tricle. The latter is common to the cerebellum and the beginning of the dorsal medulla; it is furnished in its anterior and posterior region with a process of the cerebellum which, because of its windings, we compare to a worm born in wood [21].

The gaze of the dissector encounters no peculiar body in this fourth ventricle such as may be found in the three former ventricles [22]. For in the right ventricle (as also in the left), a rather noteworthy branch of the soporal artery [23] extends through its inferior and posterior region, destined to form a plexus or network which we compare to the outermost covering of the fetus [24]. It is constituted of that branch of the artery and a portion of its vein which fashions a gland shaped like a pine nut [25] and supported by the testes of the overlying cerebrum. It is led through the third ventricle of the cerebrum from the end of the fourth sinus of the dural membrane [26], which extends along the length of the cerebellum. From that sinus, as from a press [27], it receives the content of the vein and the artery. Then it divides into two portions, of which one extends to the right, the other to the left, ventricle, and with the branches of the arteries passing to the same place it forms the plexus just mentioned in each ventricle.

From the vital spirit adapted in this plexus to the functions of the brain and from the air which we draw to the ventricles of the brain when we breathe in, the inborn force of the brain's substance creates the animal spirit, of which the brain makes use partly for the functions of the chief portion of the mind. Part of it the brain transmits by means of the nerves growing forth from itself [28] to the organs which stand in need of the animal spirit. (These are chiefly the instruments of the sense and of voluntary movement [29].) A not inconsiderable part of the animal spirit spreads from the third ventricle under the testes of the brain into the ventricle common to the cerebellum and the dorsal medulla. This is subsequently distributed to all the nerves drawing their origin from the dorsal medulla [30].

From the middle of the base of the brain a long, rounded process arises on either side and is led along the base of the brain anteriorly; it lies on the corresponding sinus of the eighth bone of the skull and is the organ of smell. Since it does not pass from the cavity of the skull, it has not been given the name of a nerve by professors of dissection [31].

The first of the seven pairs of nerves which are ascribed to the brain has its origin from the base of the brain a little behind the process corresponding to the substance of nerves, of which mention has just been made. It constitutes the optic nerves, which end in the tunic of the eye similar to a net in appearance [32]. The eye has a crystalline humor in its center, the anterior region of which is covered by a tunic corresponding to the very thinnest membrane of an onion [33]. The posterior region of this humor is held in by a vitreous humor, around the posterior aspect of which runs an involucrum

69

corresponding to the substance of the brain and in which the substance of the visual nerve is dissipated [32]. The thin membrane of the brain covers up the visual nerve and extends to a tunic very similar to the skin of a grape [34]. Although it completely surrounds the eye, it is seen to be perforated in its anterior region with an opening which we call the pupil. The dural membrane of the brain surrounds the visual nerve and ends in the hard tunic of the eye [35], which covers up the eye completely; in the anterior region of the eye it is transparent in the manner of horn [36]. This is circumscribed by the iris, the larger circle of the eye, to which clings the white of the eye or the tunic growing in the anterior region of the eye. Between this cornea and the anterior region of the crystalline humor is the aqueous humor, which is separated from the vitreous by a certain thin tunic bearing cilia which is joined in circular fashion to the crystalline humor; this tunic arises from the uvea [37].

The second pair of nerves serves for moving the muscles of the eye [38]. The third pair grows out on either side with two roots separate from each other [39]; it sends the lesser root by certain little openings to the skin of the forehead [40], to the upper jaw and upper lip, to the breadth of the nostrils [41], and to the muscles which raise the lower jaw [42]. The larger root is sent to the tongue; by the function of this root the tongue is made the instrument of taste [43]. Also, from this root a branch twisted inward in the manner of a tendril is sent to the muscles just mentioned, and another branch to the upper teeth [44], another to the lower jaw and to the teeth fixed in it, and finally to the lower lip [45].

The fourth pair of nerves ends in the covering of the palate [46]. The fifth pair proceeds with a double root also [47], as does the third pair, and distributes the lesser root to the muscles which raise the lower jaw [48] but sends the thicker root to the organ of hearing [49]; from this root it also sends two offshoots extending through different openings to the muscles just mentioned.

The sixth pair [50] is increased by a portion of the seventh pair beyond the little branches sent out from itself to certain muscles in the neck and to the larynx [51]. Near the top of the breastbone it sends certain offshoots to the muscles issuing thence [52]. To the roots of the ribs it sends down a branch [53] which is well distributed to the organs which have charge of making the blood. Thus far, each nerve of the sixth pair is equally distributed. The right nerve separately turns back a portion of itself to the artery which extends to the right arm, from which the nerve is carried up along the right side of the rough artery and proceeds to the larynx; for this reason it is called the recurrent nerve [54]. The remaining part descends from this right nerve and sends little branches to the right part of the lung and to the covering of the heart [55]; this part finally joins the esophagus after passing through the septum and giving many offshoots to the left side of the superior orifice of the stomach [56]. The left nerve turns back those portions of itself which constitute the recurrent nerve of the left side to the trunk of the great artery extending to the back [57]. From the nerve of this side a slender offshoot belonging to the heart is distributed [58]; that which remains of it is interwoven in the right side of the superior orifice of the stomach and sends a little branch as far as the liver along the upper part of the stomach. The seventh pair, in addition to the fact that it increases the sixth pair considerably, ends especially in the muscles of the larynx and the tongue [59].

The nerves which draw their origin from the dorsal medulla contained in the vertebrae include thirty pairs [60]. Of these, seven are dedicated to the cervical vertebrae, twelve to the thoracic vertebrae, five to the lumbar vertebrae, six to the os sacrum; no nerve, however, springs from the coccyx. The pairs which begin to descend from the cervical vertebrae are distributed to the muscles which rise very close to these pairs; one nerve, belonging to the transverse septum, is formed on either side from the fourth, fifth, and sixth offshoots of the pairs [61]. Then from the fifth, sixth, seventh, and next from the eighth and ninth, or from the first and second pairs of the thorax, a varied plexus of nerves arises [62] from which six nerves sprout forth into the arm, in addition to various offshoots dispersed to the hollow and to the gibbous part of the scapula.

The first nerve which seeks the arm [63] sends from its own offshoots to the muscle which raises the arm a very slender shoot to the skin covering the external region of the arm. The second nerve enters the arm through the axilla [64]; it sends little branches to the first muscle of those which flex the elbow. It imparts a notable portion of itself [65] to the third nerve which approaches the arm, and itself [66], hastening to the elbow, sends a branch to the first muscle which leads the radius in supination; it runs under the skin and, divided into various offshoots, it intermingles with the skin of the superior and internal region of the forearm as far as the hand [67]. The third nerve also descends through the axilla [68]; it sends little branches to the skin of the anterior region of the arm. Increased by a portion of the second nerve and communicating an offshoot to the posterior muscle of those which flex the elbow, it hastens to the forearm through the anterior aspect of the inner tubercle of the humerus [69]. Together with the fifth nerve, it sends shoots to the muscles which draw their origin hence; extending along the radius and led into the palm of the hand, it sends two offshoots to the internal aspect of the thumb, the same number also to the index finger, and one offshoot only to the external side of the inner [palmar] region of the middle finger; not infrequently, however, it offers two little branches to the middle finger and one to the ring finger [70]. The fourth nerve of the arm is much thicker than the rest [71]; sending branches to the muscles which extend the elbow, it enters through the armpit. First sending two offshoots to the skin [72], it hastens along the posterior aspect of the arm to the external tubercle of the humerus. The nerve placed at this external region of the elbow joint distributes a branch to the skin of the external region of the forearm as far as the wrist [73], and soon dividing into two trunks [74], it gives offshoots to the muscles growing forth from the external tubercle of the humerus. To the ulna it extends one trunk [75]; from this trunk little branches are distributed to the muscles arising from the external region of the ulna; the trunk itself ends near the root of the wrist. The upper trunk [76] reaches to the radius; except for the slender shoots which it offers to the overlying muscles, it seeks the wrist and imparts two little branches to the external region of the thumb, two likewise to the external region of the index finger, and one to the internal side of the middle finger. The fifth nerve [77] lies hidden in the armpit, nearest to the artery of the arm. It sends out no offshoots from itself to the arm. It extends into the forearm along the posterior region of the internal tubercle of the humerus [78] and, together with the third nerve, communicates branches also to the muscles growing forth here. It runs along the ulna to the wrist, sending out in the middle of its course a branch which dwindles by giving off two shoots to the exterior [dorsal] region of the little finger, two likewise to the same region of the ring finger, and one shoot to the external side of the exterior region of the middle finger [79]. Whatever portion of the fifth nerve reaches the internal region of the wrist offers small branches to the interior [palmar] region of the little finger as well as of the ring and middle fingers [80]. The sixth nerve is especially slender [81]; it is led along the internal aspect of the arm under the skin, and in its progress it extends some little branches to the skin. It reaches to the forearm, in the skin of which it is disseminated with numerous shoots along the ulna as far as the wrist.

From the nerves which spring forth from the thoracic vertebrae, with the exception of branches which are stretched backward to the spines of the vertebrae and hence to the muscles which draw their origin from them [82], single intervals of the ribs appropriate to themselves single branches stretch-ing in circular fashion to the middle of the chest and abdomen and dispers-ing shoots to the muscles laid down in the thorax, to the muscles of the abdomen, and finally to the skin. Small portions are distributed to these intervals from the intercostal nerves, which augment the offshoots of the sixth pair of nerves of the brain extending to the roots of the ribs [83].

The distribution of the nerves proceeding from the lumbar vertebrae cor-responds in large part to the nerves of the thorax, for the former send branches backward and these ascend along the ilia in a circular fashion to the middle of the abdomen, furnishing little branches to the contiguous muscles and to the skin. From the first pair of these [84], very small offshoots extend to the testes with the seminal arteries. The nerves which proceed to the thigh [85] take their origin from the four lowest pairs, although the greatest nerve [86] of all arises from the first four pairs, of the nerves of the os sacrum. The first pair of sacral nerves, like those of the thoracic and the lumbar region, descends from the vertebrae. The five lowest pairs of the os sacrum do not spring from the sides but issue forth with one root ante-riorly and another root posteriorly [87]; the posterior roots are dispersed to the muscles which attach to the bones of the sacrum and ilium and to the skin [88]. The anterior branch of the first pair, together with the anterior roots of the three succeeding pairs, constitutes the [sciatic] nerve just mentioned. The roots of the lowest pairs disappear into the bladder, anus, and penis, or in women into the neck of the uterus [vagina] and the hillocks of the pudenda [89].

Of the four nerves which proceed to the thigh, the first is led along the sixth muscle which moves the femur and sends a branch into the external aspect of the skin of the thigh [90]; it disappears in the muscles which occupy the external side of the femur. The second enters the thigh along with the larger vein and artery of the thigh [91]; it soon sends a branch through the internal aspect of the thigh, the knee, and the leg as far as the foot and the ends of the toes. It descends under the skin, together with a vein which has been mentioned as creeping forward here, and sends little branches here and there [92]. That which is left of the second nerve ends in the muscles covering the anterior region of the femur. The third nerve creeps through the foramen of the pubic bone and offers branches to the muscles which occupy it [93]. Distributed here and there, it sends an offshoot to the internal aspect of the skin of the thigh. The rest of that nerve is divided and sent to the muscles located in the internal region of the thigh. The fourth nerve [94] is easily the thickest of all the nerves of the body which are composed of more than one nerve; it is led into the posterior part of the thigh where the bone of the coxendix splits away from the sacrum. To the posterior aspect of the skin of the thigh, it sends a branch which presently ends a little below the middle of the femur [95]. Another branch propagated from the fourth nerve is offered to the lower region [96]. It also sends off-shoots to the muscles arising from the lowest posterior region of the coxen-dix bone [97], as also to the muscles which arise from the inferior heads of the femur. Then in the popliteal region it is divided into two trunks. The more slender outer one is carried to the fibula [98], from which a branch creeps to the external aspect of the skin of the leg as far as the little toe. Another branch is sent to the anterior aspect of the skin of the leg. The rest of that trunk extends to the fibula, where lies the origin of the seventh and eighth muscles of those which move the foot. The larger inner trunk [99] gives off a branch into the internal aspect of the skin of the leg and likewise to the skin of the calf as far as the heel. The trunk itself is hidden in the muscles which form the calf and sends a branch along the membranous ligament where the fibula is joined to the tibia [100]. The branch is here hidden in the muscles which occupy the anterior region of the leg; finally it reaches the upper part of the foot and is hidden there in the toes. The greater portion of that larger trunk [101] runs down along the posterior part of the leg and sends some offshoots here and there to the muscles. It approaches the lower part of the foot between the heel and the inner malleolus; it offers quite small offshoots to the muscles lying therein and communicates two offshoots to the inferior aspect of each of the toes.

In this manner, indeed, the great Creator of things has fashioned our body not only that it may live but that it may, subject to corruption, live properly. But what means he has fashioned for the continuation of our species and how he has joined them with the organs of nutrition far from the seat of the senses and of the reason [102], I shall now relate below, summarily and in so far as this enumeration of the parts of the human fabric admits.

GENERAL INTERPRETATION

THE brain, the source of activity and the seat of consciousness, is seated in the spherical, "most perfectly formed" portion of the body, the head. In addition to its protection within the skull, it is surrounded by membranes, the outer tough and the inner thin and more fragile. These membranes are not named by Vesalius but correspond to our "dura mater" and "piarachnoid." These terms may be referred to a philosophic concept employed by the Arabic anatomists: the membranes which cover a given structure are those in which the structure was originally formed. Our modern terminology utilizes this in naming the layers of the uterine wall "endometrium" and "myometrium"; both the term "mater" and the combining form "metrium" are cognates of our word "mother" and have equivalent value in the Arabic concept; the idea was carried across with the translation of the Arabic term. Mondino does not use the idea, although the names are employed; he does point out the function of the pia mater in nourishing the brain substance. It is noteworthy that Vesalius not only omits these terms but also avoids using the term "involucrum," which carries somewhat the opposite meaning of a membrane which has been derived from the structure it encloses.

Among the products of the brain's activity is phlegm, one of the four humors. It passes down the infundibular ("funnel-shaped") recess of the third ventricle and is "instilled" by the pituitary gland, passing thence to the palate and nostrils through inadequately described foramina.

The circulation of the brain is somewhat confused; the arterial circle is not recognized, and the dural sinuses perform both arterial and venous functions for the brain substance. The carotid artery sends a branch to make up the choroid plexus. This structure is remarked especially in the two lateral ventricles and the third ventricle but not the fourth ventricle (all of which spaces in the brain are described in a manner conformable with modern anatomy; this is not the case in the work of Mondino). The choroid plexus is the site in which the vital spirit, carried in the arterial blood from the heart, is turned into animal spirit (*pneuma psychicon*), which is adapted to nervous function. This is accomplished by interaction with that part of the inspired air which has entered the cranium (see Chap. IV, general interpretation) and has reached the ventricles. The animal spirit thus created is distributed along the cranial nerves.

The seven cranial nerves correspond closely to those of Galen and Mondino. The only one in which data are lacking for a fairly positive identification is the fourth. It is interesting to observe here a discrepancy in the text of Mondino. In his section "Of the Nerves Arising in the Brain," he describes the fourth nerve as our "vagus" (the key to the identification is the presence of the *nervi reversivi* or recurrent laryngeal); the sixth pair "go to the palate to give it feeling." Yet earlier, in treating "Of the Trachea Arteria," the vagus is enumerated as the sixth nerve. Vesalius says, of the fourth nerve, only that it "ends in the covering of the palate." Singer, in annotating his translation of Mondino, does not follow the confused listing of the section on cranial nerves.

76

In the comparative table below, the fourth and sixth nerves of Vesalius are not presented as positive and with complete identification; the description in the *Epitome* is inadequate for this purpose. The portion ascribed to Galen is taken from Charles Singer's *The Evolution of Anatomy* (1925), page 56; that of Mondino is from the same author's translation and annotation (note 97) of the "Anathomia," found in *Monumenta Medica,* Volume II, Part I (1925); this text was also quoted in the preceding paragraph. The development of anatomical thinking on the subject of the cranial nerves from Galen through Mondino and Vesalius to modern times will be made apparent at a glance from this table.

COMPARATIVE TABLE OF CRANIAL NERVES

MODERN USAGE	GALEN	MONDINO	VESALIUS
I. Olfactory	Not regarded as separate nerves	Not regarded as separate nerves	Not regarded as separate nerves
II. Optic	"The soft nerves of the eyes"	First pair	First pair
III. Oculomotor	"The nerves moving both eyes"	Second pair	Second pair
IV. Trochlear	Not described	Not mentioned	Included with "second pair"
V. Trigeminal	"Third pair of nerves"	Third pair	Third pair, which has two roots
	"Fourth pair of nerves"	Fourth pair	Fourth pair
VI. Abducens	United with second (II)	Not mentioned	Included with second pair
VII. Facial	"Fifth pair of nerves"	Fifth pair	Fifth pair
VIII. Auditory			
IX. Glossopharyngeal	"Sixth pair of nerves"	Sixth pair	Sixth pair; also includes sympathetic trunk (note 83) with this cranial nerve
X. Vagus			
XI. Spinal accessory			
XII. Hypoglossal	"Seventh pair of nerves"	Seventh pair	Seventh pair

1. THE average weight of the cerebellum is 140 gm; average weight of the encephalon is slightly less than 1,400 gm, although varying from 1,100 to 1,700 gm.

2. The dura mater.

3. In the cranial vault the endocranium (outer layer of the dura, which is here separable into two layers) is continuous through the sutures with the pericranium (periosteum, the "skull's own involucrum").

4. The subdural space is only a potential one.

5. The falx cerebri.

6. The tentorium cerebelli.

7. Just which dural sinuses are meant is not clear; probably we are dealing with the right and left transverse sinuses, the superior sagittal, and the straight sinus (and possibly in this numerical order). Some readers may need to be reminded that the term "sinus" in anatomy is used in many different senses and does not necessarily refer to the accessory nasal air cavities. Here the meaning is an intracranial space filled with venous blood. The reference to veins and arteries is more likely the prevailing concept of blood circulation; in only one place is a major artery (the internal carotid) included within the walls of a sinus (the cavernous).

8. The piarachnoid.

9. And serving much the same purpose—that of obtaining an increased area within a limited space.

10. The right lateral ventricle, with its inferior horn.

11. The septum lucidum.

12. The "callous body." By this name a sharp distinction is made between the texture of this white fiber tract and the gray cellular tissue composing the cortex.

13. The fornix; its old name (corresponding to "tortoise shell") was *testudo cerebri;* the parenthetic "vault" is an alternative translation.

14. Like a man standing with a broad base; legs wide apart and diverging from the trunk.

15. The third ventricle.

16. The infundibular recess.

17. The pituitary gland and the sphenoid bone (*pituita,* "phlegm").

18. Through perforations in the sphenoid, not through the lamina cribrosa ethmoidalis; see Chap. I, note 9.

19. The aqueduct of Sylvius.

20. *Nates et testes,* the superior and inferior colliculi (corpora quadrigemina; from a fancied resemblance to the appearance of the buttocks and scrotum of a man bent over, seen from the posterior aspect. It is appropriate here to remember that much of the nomenclature of the brain was established at a time when brains were not subjected to the action of a hardening solution such as formalin. The soft mass assumed shapes which might lend themselves to comparisons not so obvious in the firmer brains with which the present-day dissector is more familiar; hence the cornu Ammonis, calamus scriptorius, hippocampus, obex, clava, and many others. This comment is not to disparage the terms, for the similarities can be observed in formalinized tissue, but to emphasize that there may have been even more stimulus to the imagination of the early dissector. The *nates et testes* comparison is by no means inept.

21. The vermis of the cerebellum; a "worm born in wood" is probably a grub, the larval form of a beetle.

22. The amount of choroid plexus in the fourth ventricle is small.

23. Soporal artery: the artery of sleep (the internal carotid). Carotid is derived from καρος, "stupor," thought to be produced by pressing on the carotids. Professor J. B. de C. M. Saunders suspects an even older, onomatopoeic derivation. *Carus,* or *sopor caroticus,* was the final degree of unconsciousness and insensibility (sopor, coma, lethargia, and carus).

24. The anterior choroidal artery, see Chap. IV, note 49. The reference is to the choroid plexus, named from the fetal membrane *chorion.*

25. The pineal gland.

26. The straight sinus. The "vein of Galen," or great cerebral vein, is the forward continuation. The omission of the eponymic is probably without significance; the only man named in the entire *Epitome* proper is Plato. This is definitely not the situation in the *Fabrica.*

27. "Press": not the *torcular Herophili,* the "wine press of Herophilus" (or sinus confluens); while the figure of speech may be the same, the place under description is at the anterior end of the straight sinus, while the torcular is posterior.

28. The history of the concepts of the nerve impulse has shown two main ideas present at various times; nerves have been thought either to be somehow hollow (or pithy), transmitting a vital spirit which obeys the laws of fluids, or (on the other hand) solid, conveying a less tangible and material influence.

29. A clear distinction between sensory and motor functions.

30. The anatomical specificity surely implies some tangible carrier of this animal spirit such as the cerebrospinal fluid. Even if this be true, it does not mean that Vesalius fully accepted either the fluid or the less corporeal concept of nerve impulse propagation. A translation of that portion of the *Fabrica* dealing with this point may be read in Lecture X of Sir M. Foster's *Lectures on the History of Physiology* (Cambridge University Press, 1901), and there Vesalius refuses to commit himself, saying "we will not too anxiously discuss" the question.

31. The olfactory tract; our "first cranial nerve" which is, of course, neither first nor nerve. Vesalius explains that he does not give it the status of a nerve because it does not emerge from the cranium. As previously noted (see note 9, Chap. I), he does not describe emergence of olfactory filaments through the lamina cribrosa ethmoidalis.

32. Optic nerve; retina (from *rete*, "a net").

33. The lens and the anterior portion of its capsule.

34. The choroid tunic. "The skin of a grape" probably indicates its dark pigmentation. Note that the old names, "tunica aciniformis" or "rhagoides" or "uvea," all suggest a grape.

35. Tunica sclera.

36. Cornea (*cornu*, "horn").

37. Zonula and orbiculus ciliaris.

38. The oculomotor (our third cranial) nerve.

39. The trigeminal (our fifth cranial) nerve. This text is difficult to reconcile with the two roots at present distinguished: a large sensory and a smaller motor root ("nervus masticatorius").

40. Ophthalmic nerve.

41. Maxillary nerve, including nasal rami.

42. Masticator nerve.

43. Lingual nerve (not, of course, the nerve of taste as Vesalius thought, but the general somatic sensory innervation of the tongue; the chorda tympani runs with it in its distal portion and carries the taste fibers).

44. Superior alveolar branches of maxillary.

45. Inferior alveolar nerve, with the mental ramus.

46. The brevity of the text does not allow adequate identification of this nerve; it is probably the group of palatine nerves, branches of the trigeminal, and corresponds to the "fourth nerve" of Galen and Mondino.

47. Facial and auditory nerves (our seventh and eighth cranial).

48. A part of the facial nerve going to the mimetic muscles (here the branch to buccinator muscle is emphasized; see Chap. II, note 15, for inclusion of buccinator with the muscles of mastication).

49. The auditory nerve, with part of the facial nerve passing through the stylomastoid foramen and hiatus of the facial canal.

50. Vagus nerve, our tenth cranial, and spinal accessory (or eleventh). See general interpretation.

51. Superior laryngeal nerve.

52. Superior cardiac nerves(?). They are given off at about this level and pass near the trachea and surrounding muscles.

53. Sympathetic trunk.

54. The right recurrent (inferior) laryngeal nerve, passing under the right subclavian artery.

55. Pulmonary branches and right inferior cardiac branches.

56. Gastric branches.

57. Left recurrent laryngeal, passing under the arch of the aorta.

58. The left inferior cardiac branches usually arise only from the recurrent nerve.

59. The hypoglossal (our twelfth cranial) nerve, including (here) the descendens cervicalis portion of the ansa hypoglossi. It is probable that Vesalius considers this a portion of the hypoglossal nerve, thereby omitting the root which we count as first cervical (see following note).

60. We number thirty-one pairs; eight are cervical, of which the first is atypical and arises above the first cervical vertebra, the next being typical and arising below the vertebra. We count five sacral and one coccygeal nerves. In determining the origin of any nerves from the cervical region, we must add one to the Vesalian enumeration to reach the modern count.

61. No system of enumeration exactly accounts for this composition of the phrenic nerve.

62. The brachial plexus; quite definitely postfixed innervation of the arm (see Wilfred Harris' monograph *The Morphology of the Brachial Plexus,* Oxford University Press, London: Humphrey Milford, 1939).

63. The axillary nerve, with branches to the deltoid muscle and lateral brachial cutaneous nerve.

64. The lateral cord of the brachial plexus.

65. The lateral head of the median nerve.

66. The musculocutaneous nerve. The brachioradialis receives its innervation from the radial nerve; perhaps the inconstant branch to the pronator teres is meant.

67. Lateral antibrachial cutaneous nerve.

68. Median nerve.

69. Over the medial epicondyle, near the brachialis.

70. The variation pattern is well described, but the frequency is reversed; the emphasis on the inconstancy should be on the ring finger.

71. Radial nerve.

72. Posterior brachial cutaneous nerve.

73. Dorsal antibrachial cutaneous nerve.

74. Superficial and deep.

75. The deep branch, dorsal interosseous nerve.

76. The superficial branch.

77. The ulnar nerve.

78. Behind the medial epicondyle.

79. Dorsal digital branches.

80. Volar digital branches.

81. Medial antibrachial cutaneous nerve.

82. Dorsal primary divisions.

83. The sympathetic trunk and rami communicantes.

84. Ilioinguinal nerve.

85. The femoral nerve.

86. The sciatic nerve.

87. Through the anterior and posterior sacral foramina, respectively.

88. Such as branches to the multifidus muscle and the middle cluneal nerves.

89. Pudendal nerve, and other branches of pudendal plexus.

90. Lateral femoral cutaneous nerve.

91. Femoral nerve.

92. Saphenous nerve.

93. Obturator nerve.

94. The sciatic nerve, passing through the greater sciatic foramen.

95. Posterior femoral cutaneous nerve.

96. To the buttocks—*humilior;* the inferior cluneal nerves.

97. To the hamstring muscles.

98. The superficial peroneal nerve; inspection of the text will establish several minor errors in the distribution pattern of the terminal branches of the sciatic. The branches of the superficial peroneal here listed are the dorsal cutaneous (medial and lateral) and the muscular branches to peroneus longus and brevis.

99. This will, by necessity, include the tibial nerve and the deep peroneal nerve. The cutaneous branch is the medial sural cutaneous nerve, continued as the sural.

100. The deep peroneal (anterior tibial) nerve, which passes around the head of the fibula to enter the anterior compartment of the leg.

101. The tibial nerve.

102. The reason for this is dependent on the concept of the tripartite soul; see note on Plato's *Timaeus* in Chap. III, general interpretation.

CONCERNING THE ORGANS WHICH MINISTER

TO THE PROPAGATION OF THE SPECIES

CHAPTER VI

IN THE beginning, the Author of the human fabric fashioned two human beings for the conservation of the species in such a way that the male should furnish the primary principle of the infant, the female indeed should fitly conceive it and should nourish the little child arising from this principle as she would nourish some member of her own body [1] until the child should become stronger and could be given forth into the air which surrounds us. Both male and female received instruments suitable for these functions and peculiar to them alone. To these organs was imparted so great a power and attraction of delight in the generative act that the living creatures are incited by this power, and whether or not they are young or foolish or devoid of reason, they fall to the task of propagating the species not otherwise than if they were the wisest of beings.

The male possesses two testes covered by skin which is here called scrotum, surrounded with a fleshy membrane [2], and formed of a white, continuous substance quite peculiar to them. A strong membrane contains this substance, growing very close to it in circular fashion and receiving the insertion and connection of those parts which are near the testis, constituting a covering belonging to each of the testes [3]. Joined to this one is another covering proper [4] to the testis and growing forth from the peritoneum where the latter offers a path for the seminal vessels. Thence grows forth the membrane containing those vessels with the testis; it is at no point fused to the testis nor even to the seminal vessels (except where they escape from the great cavity of the peritoneum). This tunic is attached with its fleshy part only to the lower region of the testis; we consider this the muscle of the testis [5].

The seminal vessels consist of one vein and one artery on each side. The vein which seeks the right testis grows forth from the anterior region of the trunk of the vena cava below the place of origin of the veins which extend to the kidneys. But that vein which is offered to the left testis is believed to take its origin from the lower aspect of the vein which approaches the left kidney [6] for the reason that it may not carry the pure blood to the testis in the manner of the right vein but rather the serous blood, which by its salty and acrid quality may bring about an itching for the emission of the semen. Both arteries take their origin from the great artery a little below the right seminal vein; the right-hand one, climbing over on the trunk of the vena cava, joins the right vein, reaching the testis together with this vein. It is complexly intertwined with the vein before it reaches the testis and forms a body showing many dilatations [7]. This body is inserted with its base in the superior region of the testis, offering small branches to the innermost covering of the testis [8]. The body is distributed through the manifold substance of the testis which changes this benign blood and spirit by its own innate faculty into semen not otherwise than the substance of the liver changes the thick juice carried there from the intestines into blood.

The semen, when it is created, is received by a strong vessel like a worm growing in the posterior region of the testis and complexly intertwined like a tendril [9]. This vessel, round like a nerve, rises upward to the great cavity of the peritoneum along that path by which the seminal vein and artery descended [10]. Turning downward to the pubic bone, it reaches the posterior region of the bladder, to which a vessel bearing the semen from the left testis also runs. This vessel is attached to the right one [11] and together with it is inserted into the root of the neck of the bladder in the glandulous body covering the neck [12].

For the semen and for the urine there arises a common channel which is led slightly downward and again bends back upward to the joining of the pubic bones outside, lying under the bodies which constitute the penis. There issues forth on either side of the pubic bone a nerve and a round and sinewy body which is seen to be very funguslike within and full of blood [13]. United and fused together, they constitute the penis; by the aid of its substance it provides for its erection and enlargement when it is about to inject the semen into the uterus. Otherwise, when there is no need for its full length, it is flaccid and slender. For the satisfactory use of Venus, it swells in its apex in the manner of an acorn [14]; it is furnished with a skin by which it can be covered and uncovered [15].

The female possesses a uterus, dedicated to receiving the semen and to containing the fetus. The uterus lies between the bladder and the rectum and, like the bladder, is suitably formed with a fundus and neck [vagina] [16] adapted for stretching and relaxing, intertwined with loose membranes and with some fleshy fibers (by the assistance of which the uterus is voluntarily moved somewhat). It is joined at its sides to the peritoneum [17], just as the mesentery contains the intestines. The shape of its fundus is not completely round but flattened in front and behind, obtuse above, and showing two blunt angles (one on each side) which resemble the immature horns on the foreheads of calves [18]. In the fundus of the uterus is a simple sinus [19] corresponding very closely to the shape of the fundus; ending in an orifice and constricting and relaxing itself by a natural force alone and not by the conscious will of the female, it projects like the glans penis into the cavity of the neck of the uterus [vagina]. The fundus of the uterus consists of a simple intrinsic tunic notably thick in nonpregnant women so that it can be stretched to a remarkable extent in the uterus of women who are with child. Another tunic is drawn over this one; it takes its origin from the peritoneum. The neck of the uterus [vagina] is round and smooth; in the nongravid uterus it is not particularly distended, not much smaller than the fundus itself. It receives the insertion of the neck of the bladder [urethra] and is furnished at its orifice with leathery pieces of flesh and hillocks or wings [20].

On each side of the uterus lies one testis to which vessels extend in exactly the same way as in males [21]. Here, however, it happens that only the middle part of the seminal vein and artery is sent to the testis; the other part is interwoven in the fundus of the uterus [22]. The vessel carries a thin and very scanty and watery semen from the female testis; it is inserted into the obtuse angle of the side of the uterus. Veins and likewise arteries in a very rich series interweave the uterus, in addition to those mentioned before; they issue from those distributions of the vessels which are formed below the junction of the os sacrum and the lowest lumbar vertebra [23]. These vessels serve for nourishing the fetus and for the re-creation of the innate heat.

The fetus, contained in the uterus, is covered with three involucra. One of these is commonly called the secundine ["afterbirth"], because it surrounds only the fetus like a wide belt and is notably thick and blackish like the spleen [24]; this is joined to the uterus and receives the vessels extending thereto. By means of it, those vessels which are presently gathered within it are inserted into the umbilicus by two veins and the same number of arteries, and at last with one vein to the liver [25]. Two arteries [26] extend to the offshoots of the great artery as these are about to descend through the apertures of the pubic bones. The second envelope, collecting the fetal urine between itself and the third envelope, is a membrane embracing the entire fetus in the image of a sausage [27]. The urine is carried by a special channel [28] from the upper aspect of the bladder into this cavity so that, the urine of the fetus being contained within this membrane, the fetus is not injured by its acidity. The third envelope is a very thin membrane and is for this reason also called *agnina,* lamblike, by the masters of dissection [29]. It covers the fetus very closely; finally it keeps the perspiration of the fetus [30] between itself and the skin of the fetus, smeared, as it were, with a yellowish scum [31].

When the fetus is given forth into the light of day, it sucks the milk as its own nourishment from the breasts, untaught by anyone. The breasts have their location in the chest and are furnished with nipples; they are built up of a glandulous material which, by an innate force, converts the blood brought to them by the veins into milk.

THE END OF THE ENUMERATION OF ALL THE PARTS WHICH OCCUR
in the fabric of the human body, as far as possible most briefly and in their
completeness set down; the following pages contain the delineation
of these same parts, to be examined in the order we
prescribed at the outset.

GENERAL INTERPRETATION

VESALIUS' ideas on reproductive physiology are Aristotelian in origin. The substances from which the embryo is formed are "genital semen" and "menstrual blood." The former is formed by the testes from the blood coming to them in a manner similar to that in which the liver creates blood from the food brought to it. In addition to the menstrual blood, the female also forms a thin watery semen in the ovaries which is carried to the uterus.

The chapter appears to be the least adequate of the various sections in the *Epitome*. Much anatomy is not described. The significance of the clitoris was at first missed by Vesalius and later denied (after having been pointed out by Fallopius). The vagina goes by the name of "cervix (neck) of the uterus," and the "fundus of the uterus" is equivalent to the entire uterine mass. The uterine tubes are not described. Fugitive sheets published by Vesalius only a few years before the printing of the *Epitome* represent the uterus as bearing horns. The section on fetal membranes is not written from the standpoint of human anatomy.

The *Epitome* does not describe the various male and female reproductive organs in terms of direct homology, although implications are often present. Additional data may be drawn from the manikin-plates, where the reproductive systems have been represented in such a way that homologizing is apparent. From such sources the following table of homologues in the *Epitome* is offered; it is subject, of course, to possible revision and addition by future translators of the *Fabrica*.

MALE	FEMALE
Testis	Ovary—"testis"
Epididymis	Pampiniform plexus of veins
Ductus (vas) deferens	Portion of ovarian artery
Prostate gland	Fundus of uterus
Corpora cavernosa (urethrae et penis)	Body of uterus
Glans penis	Cervix of uterus (modern terminology)
Prepuce	Vagina—"neck of the uterus"—and labia minora

1. THAT is, as if it were an organ of the mother; the independence of maternal and fetal circulation was not known.

2. The dartos tunic.

3. Tunica albuginea.

4. Tunica vaginalis proprius.

5. The cremaster muscle, lying on the internal spermatic fascia (tunica vaginalis communis).

6. See Chap. III, note 79.

7. Pampiniform plexus.

8. The septula, which radiate from the mediastinum testis to the tunica albuginea.

9. The epididymis and ductus deferens (vas deferens).

10. The inguinal canal.

11. The colliculus seminalis.

12. Prostate gland.

13. The corpora cavernosa penis, on the inferior aspect of which is placed the corpus spongiosum (corpus cavernosum urethrae) previously described. The comparison to a nerve refers to the dense fibrous sheath which surrounds the erectile tissue (see Chap. II, note 1).

14. Glans: acorn.

15. Prepuce.

16. The "neck" of the uterus does not correspond to the cervix uteri of modern terminology, although the Latin text uses these words; it is to be identified as the vagina wherever named by Vesalius. To avoid confusion with the cervix, the word has been translated as neck wherever it occurs here.

17. The broad ligaments.

18. The cornua of the uterus.

19. In contrast to the bipartite and bicornuate uteri of lower mammals.

20. The vulva; especially the labia majora and minora. See general interpretation.

21. The uterine tubes were not described by Vesalius but by his student, Fallopius. Vesalius confirmed their existence when pointed out by Fallopius; cf. Cushing, p. 193. The term "testis" in Vesalius is equivalent to our "gonad."

22. The ovarian artery, with a branch to the uterus.

23. The hypogastric arteries, especially the uterine branches.

24. The placenta. The term "secundine" is approximately the Latin equivalent for our "afterbirth," as a word in common usage. The word "placenta" was originally applicable only to the discoid type of organ characteristic of a few Mammalia including Primates (*placenta,* "a flat cake"). For reasons not evident here, Vesalius has described the zonary placenta found in Carnivora.

25. The (right) umbilical vein of the fetus; ligamentum teres of the adult.

26. The hypogastric arteries of the fetus (see Chap. IV, note 32).

27. The allantois ("sausage-like"). This is present in lower mammals but is not well developed in human beings; the error was corrected by Fallopius and acknowledged by Vesalius (see Cushing, pp. 186 and 190).

28. The urachus.

29. The amnion (ἀμνός, "lamb").

30. The amniotic fluid.

31. The vernix caseosa.

THE NAMES OF THE EXTERNAL REGIONS OR

PARTS OF THE HUMAN BODY VISIBLE WITHOUT DISSECTION

THE enumeration of the names indicating the external regions and places of man is here briefly indicated, and these may be entered with advantage in the margins of the plates showing the superficies of the male and female body. Their description, however, is concise, since my purpose is to furnish only an index to the figures presented on these plates. Almost the same names may be applied to the external regions as are used for the bones of the body and the parts lying beneath the exterior. Of these names, determined by those who more correctly employ the method of dissection,* we have already recounted the principal ones in the context of the discourse as far as our design in the *Epitome* requires.

It is the habit among teachers first to divide the entire surface of the body into large regions and then to give various names to the parts of those regions. So the Egyptian doctors used to divide the body into head (caput, κεφαλή), thorax (θῶραξ), arms (manus, χεῖρες), and legs (crura, σκέλη); just as Aristotle did, they gave the name thorax to the entire trunk (ὅλμος) of the body from the throat, neck, or clavicles to the groin and the pubis, or rather all the way to the upper part of the thighs. They did not mean by thorax, as Galen and some anatomists of the first rank did, merely the region of the body which was fenced off with ribs. Others of those who attribute the faculties of the entire body to certain regions and consider the intellect as the seat of the soul distinguish a fourfold division of the surface of the body just as did the Egyptians, but somewhat differently from the latter they first divide the trunk into two regions; they place the arms and legs in

* In writing this passage, Vesalius was upholding the more ancient, pre-Galenic anatomists who, in his opinion, derived their nomenclature from direct anatomical examination; the terminology of such observers as Galen (who, he contends, saw only the superficial aspect of the human body) is subject to criticism.

one division, those parts which, properly speaking, constitute the members (artus, κῶλα) or extremities. In the trunk of the body they make a division into two special regions, according to the two cavities therein which are obvious to dissectors. Of these, the lower one is cut off from the upper one by the intervention of the transverse septum; it contains the liver, which is the seat of the natural or nourishing faculty and the place where the blood is made, the organs which minister to the liver, and the parts which serve for the task of generation. The upper cavity holds the heart, the tinder of the agitative [pulsific] faculty, and the organs subserving the heart. The principal seat of the soul is assigned to the head; this third cavity of the body is the seat of the brain and the storehouse of the animal spirit.

Having thus cursorily divided the body, the teachers again divide the surface into separate parts as follows: the anterior part of the entire head above the eyebrows, free from hair (τρίχες) and showing certain furrows (στολίδες, ἀμαρυγαί), is called the forehead (frons, μέτωπον). Above this and verging toward the middle of the head is the sinciput (6ρέγμα). On both sides of the sinciput and above the ear (οὖς), in which is the auditory meatus (ἀκουστικὸς πόρος), lies the temple (κρόταφος, κόρσος). The middle region of the head rising toward the posterior of the sinciput is called the vertex (κορυφή), like the center of a circle which circumscribes the hair margin (περίοδος, στεφάνη, περίδρομος). Near the highest region of the muscles called tendons (τένοντες) by many and behind the vertex, the occiput (ἰνίον) is seen. These muscles are prominent on either side of the neck and show a median hollow.

The front part of the head, from the forehead down to the chin, is the face (facies, πρόσωπον). The inferior part of the forehead is delimited by projecting eyebrows covered with hair (ὄφρυες) and by the interval between them (μεσόφρυον). Below these lie the eyes (oculi, ὀφθαλμοί), covered by the upper and lower eyelids (βλέφαρα); the regions where they come together mutually are furnished with eyelashes (cilia, βλεφαρίδες), a row of erect hairs like the oars we see on galleys; these regions are somewhat cartilaginous and contain the tarsi (τάρσοι). The termini of this commissure are the angles (anguli, κάνθοι); the greater one faces toward the nose, the lesser one toward the temple. Between the separated eyelids, in

91

addition to the caruncula (ἐγκανθίς) seen in the greater angle, there appears the "white" of the eyes (candidum, λευκόν), in the middle of which two circles present themselves. Of these the broader one is the iris (ἴρις, στεφάνη) or corona, the lesser is the pupil (κόρη). The nose (nasus, ῥίς) lies between the eyes; its apertures are called the nostrils (nares, μυκτῆρες, μυξωτῆρες). The external sides of these are fitted with the pinnulae or wings (alae, πτερύγια) of the nose; the inner aspect is made up of the septum (διάφραγμα). The regions at the sides of the nose, rather prominent and reddish like an apple, are called the malae (μῆλα); some call them the cheeks (genae). The regions midway between the nose and the cheeks are by some called the concava (κοῖλα), by which name others indicate the entire region of the eyes (γλήνη) extending from the eyelids to the cheeks. That part of the face which we inflate is the bucca (γνάθος). The entire region from the eyebrows to the upper row of teeth is the upper jaw (superior maxilla, ἡ ἄνω γένυς); the remaining part, which in men is adorned with the beard (barba, πώγων), is called the lower jaw (inferior, ἡ κάτω γένυς). The anterior extremity of this is led out into the chin (mentum, γένειον), sometimes provided with a hollow (fovea, τύπος, νύμφη) lying below the redness of the lower lip (τὸ κάτω χεῖλος). The region of the upper lip (τὸ ἄνω χεῖλος) lying below the nose is endowed with a little furrow (sulculus, φίλτρον); there the mustache is seen (mustax, μύσταξ, ὑπήνη). That which is circumscribed and contained by the lips is called the mouth (os, στόμα), in which when open we see the tongue (lingua, γλῶσσα), the palate (palatum, ὑπερῶα), the uvula (γιργαρεών, σταφυλή), the teeth (dentes, ὀδόντες), the gums (gingivae, οὖλα), and the internal aspect of the fauces (φάρυγξ).

That part of the body from the head to the clavicles or as far as the thorax is called the neck (collum, τράχηλος, αὐχήν) or cervix; if we restrict the latter name to the posterior part, just as the former is applied to the place where the rough artery (and especially its upper end) can be palpated (λάρυγξ), we should use the term throat (guttur).

The ancients called the joining of the armbone with the scapula the humerus (ὦμος), and from this they called that eminence at the root of the neck and to the sides of the thorax the summus humerus (ἀκρώμιον). That part which turns directly away from it toward the throat (σφαγή) and the hollow at the root of the neck is the clavicle (κλείς). The region which extends from the humerus to the tips of the fingers is the arm (manus, χείρ). The first part of the arm extending to the nearest joint, the bend of the elbow (cubitus, ἀγκών), is called the brachium (βραχίων) and by some of the Latins* also the humerus. Under it lies the cavity called the axilla (μασχάλη) or wing, walled off with muscles which in that place many call tendons (τένοντες). The posterior region of the bend of the elbow is called the gibber (ὠλέκρανον, κορώνη). The part which extends from the gibber to the conterminous joint is the forearm (cubitus, πῆχυς), also by some of the Latins called brachium or ulna (ὠλένη). At the very end of the forearm begins the hand (summa manus, ἀκρόχειρ), of which a part extends from the forearm to the roots of the four fingers, divided into two regions; the region closer to the forearm is the wrist (brachial, κάρπος), the other is the postbrachial (μετακάρπος), which from the likeness of its construction to the chest is also called chest (pectus, στῆθος) and, by some, palm (palma). The inner aspect of the latter is hollowed, fenced off by various small mounds, and crisscrossed by many lines; this region is the vola (θέναρ). The remainder of the hand forms the fingers (digiti, δάκτυλοι), each one divided into three parts (σκυταλίδες, φάλαγγες) placed in series, the most external being equipped with nails (ungues, ὄνυχες). The largest of these digits is the thumb (pollex, ἀντίχειρ), which is opposable to the others by its action; the one next to it is the index (λίχανος), then the middle finger or impudicus (μέσος), next to which is the medicus or ring finger (παράμεσος, ἰατρικός). The last place is occupied by the little finger (μικρός) or ear finger (auricularis).

We here call the thorax (θώραξ) that part of the trunk of the body which is fenced off with ribs (πλεῦραι) and forms the greatest part of the sides (πλευρά). The anterior part of the thorax is the chest (στῆθος), occupied by the breasts (mamillae, μαστοί), with the nipples (papillae, θηλή) in the middle surrounded by a darkish circle (φῶς). The remaining anterior region of the trunk forms the abdomen (ὑπογάστριον); that portion

---

* While the apparent meaning is "Roman anatomists and physicians," the term probably refers to those who adhere to the Latin terminology rather than use the preferred Greek stem.

of it which is nearest the cartilage of the breastbone and the lower cartilages of the ribs is designated as subchondral (ὑποχόνδρια) as the viscera are surrounded by them. The place where the transverse septum is called the precordia is the region in which the septum inserts into these cartilages and thus obtains its name, although, again, others give this name also to the anterior region of the thorax. The region between the lowest ribs and the crest of the ilium (which projects more in the female than in the male) is free from bone and can be palpated; it is called the inania or flank (κενεῶνες, λαγόνες). In the middle of this region the umbilicus (ὄμφαλος) is seen, and immediately below it is the paunch (sumen, ἦτρον), the lowest region of which, nearest to the end of the trunk, is called the waterpot (aqualiculus, ἐφήβαιον, ἐπίσιον).

The pudenda or naturalia (αἰδοῖα) lie at the terminus of the trunk; this end is the pubes or pecten (ἥβη), at whose sides, in the flexures of the thigh, we place the groins (inguina, βουβῶνες). The part of the male pudenda visible without dissection is called the penis or prick (coles, ["stalk"], καυλός); the summit of it is thicker than the rest of its length and forms the glans (βάλανος), in the middle of which appears the channel common to the urine and the semen. The covering of the glans penis is the prepuce (praeputium, ποσθή), although others give this name to the entire summit of the penis. A prominent line in the semblance of a suture in the covering of the glans and in the rest of the skin to the anus we call the suture (ῥαφή); the entire part of the penis stretching prominently as far as the anus we call the taurus (ταῦρος). We call interfeminium (περιναῖον, διχάλα) the region between the anus and the covering of the testes (which, since it is composed of skin, is called scrotum, ὄσχεον). The opening of the female pudenda, which is the orifice of the neck of the uterus [vagina], is called the sinus (κόλπος, κτείς); on both sides of the sinus project wings and hillocks (πτερυγώματα), and on the summit of the sinus appears a fleshy skin (νυμφή). The orifice of the intestinum rectum which appears in the region is called annulus from its shape (δακτύλιος) and stricture (σφιγκτήρ) from its function.

The posterior part of the trunk of the body is called either back (dorsum, νῶτον) or tergum; its sides in the higher posterior aspect of the thorax are

94

composed of the scapulae (ὠμοπλάται). The mid-region between them (μεσόφρενον) and the part of the back extending hence to the lowest ribs, or where the back protrudes most in flexion (ἄχνηστις), is ascribed to the thorax and is situated behind the transverse septum. The region extending as far as the buttocks comprises the loins (lumbi, ὀσφύς). The buttocks (nates, γλουτοί) are a rounded and fleshy region occupying the dorsal aspect of the iliac bones; in the midst of the buttocks the posterior processes of the os sacrum and the coccyx, subcutaneous, extend to the anus.

Where the joint of the femur appears, there protrudes the great rotator (τροχαντήρ); here is also the coxendix (ἴσχιον) or coxa, which name others assign to the femur (μῆρον), extending from the groin to the knee (γόνυ). The posterior region and the joint of the latter are called the ham (poples, ἰγνύς). The leg (κνήμη) follows from the knee to the next joint or the beginning of the foot; some call the leg the crus, while others use the same name to designate both leg and thigh. The anterior region of the leg (ἀντικνήμιον, κρέα) is bony to the touch; the posterior region of it, where the belly is seen, is fleshy and is called venter or the calf (sura, γαστροκνήμη). The tuberosities on either side at the end of the leg are, like bones, obvious and are called malleoli (σφυρά); not so is the talus (ἀστράγαλος) which, unlike these, is buried. The posterior region of the foot, projecting backward beyond the straight line of the tibia, is called the heel (calx, πτέρνα). The remaining regions of the surface of the foot take their names from the bones, especially the flat (tarsus, τάρσος) of the foot (πεδίον) or of the chest (pectus, στῆθος); the toes, furnished with nails, follow. However, when we speak of the entire foot, the lowest region on which we walk is usually called the sole (planta) or vestigium (τέλμα); its inner side is the hollow (concavum, κοῖλον), but the superior aspect is the tarsus.

THE GREEK MARGINAL NOTES OF VESALIUS
(218 ITEMS)

CHAPTER I

	MARGIN	TEXT	TRANSLATION
a	ὁμοιομερεῖς	similares	similar in temperament and elements
b	ὀστοῦν	os	bone
c	χόνδρος	cartilago	cartilage
d	σύνδεσμος	ligamentum	ligament
e	ἶνες	fibrae	fibers (muscle)
f	ὑμήν	membrana	membrane
g	σάρξ	caro	flesh
h	πιμελή	adeps	fat
i	ἀνομοιομερεῖς, uel ὀργανικαί	dissimilares	dissimilar in temperament and elements
k	φλέψ	uena	vein
l	ἀρτηρία	arteria	artery
m	νεῦρος	neruus	nerve
n	μῦς	musculus	muscle
o	δάκτυλος	digitus	finger
p	κράνιον	calvaria cerebri	skull
q	ὀστοῦν μετώπου	in fronte . . . unum tantum (os)	frontal bone
r	ἰνίου	in occipitio unum os	occipital bone (inion)
s	βρέγματος, uel κορυφῆς	in uertice duo (ossa)	parietal bones
t	κροτάφου	tempus unum (os)	temporal bone
u	μαστοειδής	unus (processus) uberis	mastoid process
x	στυλοειδής, uel βελονοειδής, aut πλῆκτρον	stylum, aut acum, aut galli calcar	styloid process
y	ζύγωμα	iugale os	zygomatic processes
z	λιθοειδής	quam praeruptae rupi . . .	(see note 6)
a	σφηνοειδές, uel πολύμορφον	in capitis basi os	sphenoid bone

96

	MARGIN	TEXT	TRANSLATION
b	πτεριγοειδεῖς [sic]	processus uespertilionum alis simillimos	projections very much like the wings of a bat (pterygoid)
c	ἠθμειδές uel σφογγοειδές	os in narium summo	ethmoid bone
d	διάφραγμα	septum	septum
e	ἡ ἄνω γνάθος	superior maxilla	upper jaw
f	κόγχος	sedis oculi regione	orbit
g	πτερά	narium alae	wings of the nostrils
h	στεφανιαῖα	coronalis sutura	coronal suture
i	λαμβδοειδής	sutura a Λ similitudine nomen. . . .	lambdoid suture
k	ὀβολιαῖα [sic]	sagittalis sutura	sagittal suture
l	λεπιδοειδῆ προσκολλήματα	squamosae conglutinationes	squamous conglutinations
m	ἡ κάτω γνάθος	inferior maxilla	lower jaw
n	ὀδόντες	dentes	teeth
o	τέμνοντες	incisorios	incisors
p	κυνόδοντες	caninos	canines
q	μυλίται	molares	molars
r	φάτνια	praesepolis	tooth sockets
s	φάρυγγες	faucibus	pharynx
t	ὑοειδές	os v imaginem exprimens	hyoid bone
u	λάρυγξ	laryngem	larynx
x	θυρεοειδής	scutum	thyroid cartilage
y	ἀνώνυμος	cartilago nomine destituitur, hinc . . . . id inueniens	cartilago innominata, the cricoid cartilage
z	ἀρηταινοειδής [sic]	vasorum orificio respondet	arytenoid cartilage
a	γλωττίς	lingulam	little reed: glottis or lingula; vocal apparatus
b	ἐπιγλωττίς	operculum	epiglottis
c	σιγμοειδεῖς	C imaginem exprimentibus	uncial capital sigma
d	ῥάχις	dorsum medulla	dorsal medulla, the spinal cord
e	αὐχήν, aut τράχηλος	collum	neck
f	θῶραξ	thoracem	thorax
g	ὀσφύς	lumbos	loins

97

THE EPITOME OF ANDREAS VESALIUS

	MARGIN	TEXT	TRANSLATION
h	ἱερὸν ὀστοῦν	sacrum os	sacrum os (holy bone), in the shape of a cross
i	κόκκυξ	coccyx	coccyx or cuculus
k	σπόνδυλος	uertebras	vertebrae
l	ἀνάνθη	spina	spine
m	ὀδοντοειδής	processus instar canini dentis	process like a canine tooth (see note 33)
n	στέρνον	pectoris os	breastbone
o	στῆθος	anteriorem sedem thoracis	anterior region of the thorax
p	θῶραξ	thorax	thorax
q	πλεῦραι	costis	ribs
r	ἀληθεῖς	c. uerarum	true ribs
s	νόθαι	c. spuriae	false ribs
t	ξιφοειδής	manubrium cartilaginem	xiphoid process
u	σφαγή	iugulum	throat, root of the neck
x	κλείς	clauicula	clavicle
y	ὠμοπλάτη	scapula	shoulder blade
z	ἀκρώμιον	summus humerus	acromion process
a	ἀγχυροειδής, uel σιγμοειδής	interior scapulae processus	coracoid or anchor process (see note 41)
b	βραχίων	brachii os	bone of the arm, or humerus
c	πῆχυς	duo inibi ossa	cubitus
d	κερκίς	radius	radius
e	πῆχυς uel ὠλένη	ulna	ulna
f	ὠλέκρανον	gibberum	gibber, the olecranon process
g	στυλοειδής	qui a styli forma nomen	styloid process
h	κάρπος	brachiale	wrist, or carpus
i	μετακάρπιον, aut στῆθος	postbrachialis	postbrachial, the metacarpals
k	φάλαγγες, aut σκυταλίδες	tribus ossibus	phalanges of the digits, or three bones
l	σησαμοειδῆ	ossicula	sesamoid bones
m	λαγόνος ὀστοῦν	ilium os	hipbone
n	ἰσχίου	coxendicis	os coxae
o	ἥβης	pubis os	pubic bone
p	μῆρον	femoris os	femur
q	γλουτός, et μέγας τροχαντήρ	natem et magnum rotatorem	rump and great rotator, the greater trochanter
r	μικρὸς τροχαντήρ	processum	minor or interior rotator, the lesser trochanter
s	κνήμη	tibia	tibia

MARGIN	TEXT	TRANSLATION
t περόνη	exteriori interim osse	fibula
u μύλη,	os rotundum. . . . mola	a round bone called mola
uel ἐπιγονατίς	patellaue	(millstone), or patella;
		(kneecap)
x σφυρά	malleoli	malleoli
y ἀστράγαλος	talum	talus
z πτέρνη	calcis os	heel, especially the cal‑
		caneum
a σκαφοειδές	ossis navicularis	navicular bone
b τάρσος	tarsi	tarsus, including the navicu‑
		lar, the three cuniforms,
		and the cuboid bones
c χαλκοειδῆ	ossa tria	three bones
d κυβοειδές	os cubi et tesserae	cuboid bone
	imaginem exprimens	
e πεδίου ὀστᾶ	ossa pedii	metatarsal bones
f ὄνυχες	ungues	nails
g τάρσοι	cartilagines	cartilages of the eyelids
		(the tarsal plates)

CHAPTER II

σύνδεσμος	ligamentum	ligament
[extra ordinem]		
a ἀπονευρώσεις	eneruationes	aponeuroses
b νεῦρος	nerui	nerves, often indicating
		sinews
c μῦς	musculus	muscle
d τένων	tendo musculi	tendon
e δέρμα	cutis	skin
f ἐπιδερμίς	cuticulam	epidermis
g ὑμὴν σαρκικός	membrana	membrane
h πιμελή	adeps	fat
i κροταφίτης	temporalis	temporal muscle of lower
		jaw
k μασσήτηρ	mansorius	masseter
i ἐπωμίς	humeri articulum	covering the shoulder joint,
	. . . . tegens	the first muscle of the arm
k δελτοειδής	Δ non absimilis	deltoid muscle (second), not
		unlike the Greek letter Δ
l μεσοπλεῦροι	exteriores musculi	intercostal muscles (see note
		33)

MARGIN	TEXT	TRANSLATION
m διάφραγμα, et φρένες, et ἡ ἄνω κοιλία	septum transversum	diaphragm
n κρεμαστήρ	testi unus musculus	cremaster
o σφιγκτήρ	alius musculus	sphincter
p ἀντικνήμιον	musculus . . . . anteriori tibiae sedi inseritur	sartorius, the first muscle of the leg (see note 56)

CHAPTER III

a οἰσόφαγος aut στόμαχος	uiam, stomachus	esophagus
b κοιλία	uentriculus	stomach
c πύλωρος	inferius orificium	lower orifice of stomach (pylorus)
d ἔντερα	intestina	intestines
e ἔκφυσις	origo	origin (of the intestines)
f δωδεκαδάκτυλον	duodenum	duodenum
g νῆστις	ieiunum	jejunum
h ἴλεον	ilium	ilium
i τύφλον	caecum	appendix vermiformis
k κῶλον	colum	colon
l ἀπευθύσμενον, uel ἄρχον	recti ac principis intestini	straight intestine, the rectum
m ἀδῆνες [sic]	glandulis	glands
n μεσεντέριον uel μεσαραῖον [sic]	mesenterium	mesentery
o ἧπαρ	iecur	liver
p πύλης φλέψ	portae uenae	portal vein
q ἐπίπλοον, uel ἐπίπλουν	omentum	the omentum [caul]
r πάγκρεας, uel καλλίκρεον	glandulis	glands, fleshy in color (pancreas)
s σπλήν	lienem	spleen
t πόροι χολήδοχοι	meatus	bile ducts and hepatic duct
u κύστις χολήδοχος	uesiculam	gall bladder
x νεφροί	renes	kidneys
y πόρος οὐρητήρ	urinarius meatus	urinary passage
z κύστις	uesica	bladder
a περιτόναιον	peritonaeo	peritoneum
b κοίλη φλέψ	cauae	vena cava
c στεφανιαῖα	uena	cardiac vein

100

MARGIN	TEXT	TRANSLATION
d ἄζυγος	uenam paris expertem	azygous vein
e ὠμιαῖα, aut κεφαλική	humerariam [*venam*]	cephalic vein
f ἡ κατὰ βάθος σφαγίτις	internam iugularem	internal jugular
g ἡ ἐπιπολῆς σφαγίτις	superficiariam iugularem	superficial jugular
h μασχαλαῖα	axillaris uenae	axillary vein
i κοινή	communem	median antibrachial vein
k μέσαι	mediis uenis	middle veins (see note 74)

CHAPTER IV

a καρδία	cor	heart
b κόλποι, siue κοιλίαι	sinus seu uentriculos	sinuses or ventricles
c ἀρτηριώδης φλέψ	arterialis uena	arterial vein, or pulmonary artery
d ἀρτηρία φλεβώδης	uenalis arteria	venous artery, or pulmonary vein
e διάφραγμα τῆς καρδίας	septo	septum
f περικάρδιον	inuolucrum cordis	pericardium
g διαφράττοντες ὑμένες	membranis	mediastinal septum
h πνεῦμων	pulmo	lung
i τραχεῖα ἀρτηρία	asperae arteriae	rough artery, trachea
k παρίσθμιαι	tonsillae	tonsils
l ὑποζωκὼς [sic] χιτών	tunicae	tunic, pleura (see note 15)
m γαργαρεών	gargareonem	uvula
n καρδίας ὦτα	cordis aures	auricles, auricular append- ages
o ἀορτή	magna arteria	great artery, or aorta
p καρῶτις	soporalem arteriam	carotid artery

CHAPTER V

a ἐγκέφαλος	animalis ac principis facultatum sedes	the seat of the animal and the principal faculty, the brain
b νωτιαῖος μελός	dorsalis medulla	dorsal medulla, spinal cord
c μυελός	ossium medulla	bone marrow
d παρεγκέφαλις	cerebellum	cerebellum
e παχεῖα μῆνιγξ	dura membrana	dura mater

101

MARGIN	TEXT	TRANSLATION
f περικράνιον	inuolucrum calvariae	pericranium (see note 3)
g λεπτὴ μῆνιγξ	tenui cerebri	pia (see notes 4, 8)
h διάφραγμα	septum	septum
i τυλλῶδες σῶμα	callosum corpus	corpus callosum
k ψαλλοειδές	parti quae instar fornicis testudinisue	fornix, arch, vault, or tortoise shell
l πύελος aut χοάνη nisi id hic descriptae glandi tribuere uisum sit	peluim	funnel or basin (infun-dibular recess), funnel, basin
m ὑπερῶον	palatum	palate
n ὀρχεῖς	testibus	testes
o γλουτοί	natibus	nates, buttocks
p σκωληκοειδής	processu	process, vermis
q χοροειδής	plexum	choroid plexus
r κωνάριον	glandis nucis pineae instar	pineal gland
s ληνός	torculari	wine press
t ἀμφιβληστροειδής	tunicam retis imagini similem	retina
u κρυσταλλοειδής, uel φακοειδής	crystallinum humorem	crystalline humor (lens)
x ἀραχνοειδής	tunica	capsule, lens
y ὑαλοειδὲς ὑγρόν	uitreo humore	vitreous humor
z ῥαγοειδής, aut ῥογοειδής	tunicam uuae folliculo similem	a tunic very similar to the skin of a grape, choroid tunic
a κόρη	pupillam	pupil
b σκληρότης aut στερεός	duram oculi tunicam	hard tunic, sclera
c κερατοειδής	pellucidam tunicam	transparent tunic, cornea
d ἶρις	iride	iris
e λευκός	alba tunica	"white" of the eye
f ὑδατώδες [sic]	aqueus humor	aqueous humor
g παλινδρομῶν	neruus recurrens	recurrent nerve

CHAPTER VI

a ὄρχεις	testes	testes
b ὄσχεον	scortum	scrotum
c ἐπιδιδυμίς	ualida membrana	a strong membrane, tunica albuginea

102

GREEK MARGINAL NOTES OF VESALIUS

	MARGIN	TEXT	TRANSLATION
d	ἐρυθροειδής, aliis δάρθος	alterum inuolucrum	another covering, tunica vaginalis proprius
e	κρεμαστήρ	testis musculum	testis muscle, cremaster
f	κιρσοειδὴς παραστάτης	corpus . . . . multas uarices exprimens	a body showing dilatations, pampiniform plexus
g	πόρος σπερματικός	uase instar uermis	a vessel like a worm, vas deferens
h	ἀδηνοειδὴς παραστάτης	glandulosum corpus	glandulous body, prostate gland
i	καυλός	penem	penis
k	βάλανος	glandis	acorn, glans
l	πόσθη	cute	skin (foreskin), prepuce
m	μήτρα	uterum	uterus
n	κέρατα	duos retusos angulos	two blunt angles, horns, cornua uteri
o	νύμφη	coriaceis carnibus	leathery pieces of flesh, labia minora
p	πτερυγώματα	colliculis alisue	wings or hills, labia majora
q	κτείς, uel κόλπος	orificium	orifice, vulva
r	χόριον	inuolucrum	envelope, afterbirth
s	ὄμφαλος	umbilicum	umbilicus (navel)
t	ἀλαντοειδής [sic]	secundum inuolucrum	second envelope, allantois (sausage-like membrane)
u	οὐραχός	meatu	channel of urachus
x	ἄμνιος	tertium inuolucrum	third envelope, amnion
y	μαστός	mamillis	breasts
z	τιτθοί	papilla	nipple

103

THE EPITOME OF
ANDREAS VESALIUS

(LATIN TEXT)

ANDREAE VESALII
BRVXELLENSIS, SCHOLAE
medicorum Patauinae professoris, suorûm de
Humani corporis fabrica librorum
EPITOME.

CVM CAESAREAE
Maiest. Galliarum Regis, ac Senatus Veneti gra
tia & priuilegio, ut in diplomatis eorundem continetur.

LECTORI.

QVOD modò datus librorum de Humani corporis fabrica Compendium, in duas partes dissectum est. quarum una sex capitibus complexa, succinctissimam omnium partium historiam complectitur: altera in pluribus tabulis earundem delineationem simul cum characterum quibus insigniuntur indice proponit. Quare tuo ipsius arbitratu nostrum ordinem: quem ex ua-
rio. d impressionis formam, compingendiq́ rationem commodissimum selegimus: tanti quanti ui sum erit pendens, aut partium descriptionem, aut designationem, notarumq́ indicem primùm ag-
gredieris, quem omnino a figuris nudas uiri mulierisq́ imagines proponentibus auspicabere, ubi externarum sedium appellationes, ceu illarum imagi num index, occurrent. Integram ossium
compagem offert figura a uiri tergo impressa: quanquam & figuræ ad uiri illius proportionè delineatæ, ac uelut musculorū tabulæ inscribtæ sextam sectionis serie essa proponant & præcipuè
earum quarta & quinta. Musculorum etenim ligamentorumq́ delineatio petenda est primùm ab illa quam è regione figuræ omnia ossa exprimentis locamus, ob idq́ prima musculorum figura
inscribitur: cui succedit quæ tertia nominetur, dein quarta & quinta. Nutritionis quæ cibo potuiq́ fit organa, & dein cor partes quæ ipsius munijs subseruiretes, s nul cum neruorum
serie proponuntur figuris nudam mulieris imaginem subsequentibus, ubi & mulierum generationis spectantur instrumenta, perinde ac ui ctum organa occurrunt n figuræ quartæ musculorum tabulæ agglutinata. Eorum quæ caluaria complectuntur imaginem, præter tabulam neruis commonstrandis paratam, musculorum f iguræ suis absolutè referunt ac primùm quidem caput primæ figuræ, dein secundæ, mox quartæ simul cum imaginibus quas ius figuræ manus amplexantur, & has si hisq iuctur quæ in sinistra quinæ
figuræ manu est obuia, at aqg inibi præter oculi partium effig iem humi decumbens. Vale, &
nostris conatibus cand de i uti offeruntur i utere.

BASILEAE.

A

SERENISSIMO PRINCIPI PHI-
LIPPO, DIVI CAROLI QVINTI IMPERATORIS

MAXIMI INVICTISSIMIQVE FILIO HAEREDI.

Andreas Vesalius   S.

TENVI hoc chartarum contextu, maxime *Princeps* PHILIPPE, *qui immortalis tui no-minis splendore decoratus, in communem studiorum usum auspicatò emittitur, humani corporis historiam enumerationis modo ita partitus sum, singulaq́ enumeraui, ut præcipua naturalis phi-losophiæ pars creaturæ omnium absolutissimæ & iuxta dignissimæ fabricam pertractans, ima-ginis cuiusdam uice operum Naturæ studiosis ob oculos uersari possit. quæ quoad fieri licet suc-cinctè, ac minus operose ea exprimat, quæ septem huius argumēti libris diffusè complexus sum, quorum hæc Epitome semita quædam, aut appendix etiam iure habebitur, capita quæ illis de-monstrantur aceruatim comprehendens, omniaq́ sic semel proponens, ut augurer eam pro tua incredibili humanitate, qua etiam tenuissimum quodq́ scripti genus alacriter amplecteris, è tuo conspectu non reijciendam, quando ut hac ineunte adhuc ætate tam uarijs uirtutibus exornata, omnium disciplinarū & artium miro amore teneris, ita quoq́ generosissimi, ac qui uniuerso orbi aliquando præficiatur digni prorsus animi tui domiciliū, uel obiter cognoscere iucūdum duces, atq́ summorum Imperatorum, Regum, Consulum exemplo turpe miserumq́ esse arbitraberis,* nos adeò uaria sectantes studia, corporis quod perpetuò circunferimus harmoniā latére, ac hominem sibijpsi penitus esse incognitum, neq́ instrumentorū tam diuinè ab immenso rerum Opifice structorum compagem perscrutari: quorum munere ea quæ potissimū, et sola prope modum admiramur, alioquin conficiuntur. *Verùm & si hac ratione conatus iste tuo omnibus admirabili iudicio forsitan non displicebit, illum tamē medicis adeò nollem reddi familiarem, ut dum prodesse conor, ac in posterum quibusdam sordidorum typographorū mancipijs ad aliorum labores ineptissimè in arctum contrahendos, suoq́ nomine emittendos ad rei literariæ perniciem natis, occasionem præripere studeo, magnopere obsim.* Nemo enim nescit, quantum dispendij in scientijs omnibus compendia conciliare soleant. Quandoquidem & si illa ad perfectam absolutamq́ rerum cognitionem uiam quandam & ratione præfigere uideantur, ac latius prolixiusq́ alibi tradita, bre-uiter uniuersimq́ contineant, hocq́ in primis nomine instar indicis & memoriæ sedis censeantur, in quam sparsim scripta ritè collocen-tur, nihilo minus compendia ideo egregiè nocent, magnamq́ literis cladem adferunt, quod illis solummodo addicti, uix quicquam aliud hodie perlegamus, etiā ijs qui se totos disciplinis manciparunt, duntaxat ad scientiarum umbram & superficiem aspirantibus, ac penitius solidè ue nihil eruentibus. Huius mali quantumuis in omnibus propemodum studijs id latè uagari constet, medicorum uulgus grauissimè insimulandum est, quod adeò ignauiter hominis partibus dignoscendis operam nauat, ut ne enumeratione quidē ad eas discendas utatur. Quum enim præter cuiusq́ partis functionem & usum, ipsius quoq́ situs, figura, magnitudo, color, substantiæ natura, ad cæteras par-tes connexus ratio, ac eius generis permulta in partium examine medico nunquam satis perspecta esse queant, quotus quisque reperitur, qui ossium, cartilaginum, musculorum & uenarum, arteriarum neruorumq́ frequentissima per totum corpus serie excur-rentium, ac uiscerum, quæ in corporis habentur cauitatibus, uel numerum modò tenet? Vt pestilentes præter eam medicos, in communis hominum uitæ exitium grassantes, qui ne unquam quidem sectioni astitére, quum in corporis cognitione nemo aliquod operæ precium præ-stiturus sit, qui non proprijs manibus, uti Ægyptiorum reges consueuerunt, sectiones obierit, & illas perinde ac simplicia medicami-na sedulò frequenterq́ uersauerit. Vnde etiam nunquam satis laudabuntur prudentissimæ Asclepiadum familiæ, liberos domi ut lectio-ne & scriptura, sic & cadauerum resectionibus exercentes, ut ita edocti, fœlici Musarum omine studijs incumberent. Cæterùm hic no-stram in Anatome totius medicæ artis basi ac fundamento perdiscenda oscitantiam, & quam nobis qui medicinæ nomen dedimus, per-necessaria sit humanarum partium notitia, haud demonstrandum assumpsi: quum cuiusq́ conscientia abundè attestetur, in morbis curan-dis eorum cognitionem primas, secundas ac tertias partes meritò sibi uendicare, hanc que in primis ab affecta sede peti, debitum quoque (præter alia) auxiliorum usum commonstrante. Quinetiam qui ueteri medicinæ nunc in plerisque gymnasijs pristino nitori propemo-dum restitutæ dedicantur, affatim intelligere incipiunt, quàm parum frigideq́ hactenus à Galeni temporibus in Anatome sudatum fue-rit: qui & si huius procerum facilè sit primarius, humanum tamen corpus nunquam aggressus est, & sunia potiùs quàm hominis ab illius fabrica innumeris sedibus uariantis partes descripsisse (ne dicam, nobis imposuisse) modò colligitur. Quòd autem ad meam attinet au-daciam, qua exiguum hoc tuaq́; indignum Maiestate munusculum, tanto patrocinio unicè commendatum, ancipitem iudiciorum aleam su-bit, nullam excusationem prætexurus sum, nisi quòd hanc falsam molam, qua maximè per studiorum rationem litare datur, tantisper meæ erga patriæ Principem singularis obseruantiæ pietatisq́ specimen esse uelim, dum thura quoq́ offerre aliquando dabitur. Patauij, Idibus Augusti. Anno à Vir-ginis partu MDXLII.

ANDREAE VESALII BRVXEL-

LENSIS, SVORVM DE HVMANI CORPORIS
FABRICA LIBRORVM EPITOME.

DE OSSIBVS, ET CARTILAGINIBVS,
seu ijs quæ corpus suffulciunt partibus. Caput I.

MNES humani corporis partes sunt aut similares, sensuue simplices, ut os, cartilago, ligamentum, fibræ, membrana, caro, adeps: aut dissimilares, instrumentariæ ue, quemadmodum uena, arteria, neruus, musculus, digitus, & reliqua totius corporis organa: quæ eo magis instrumentaria fiunt, quo pluribus simila ribus, ac dein instrumentarijs etiam partibus (ut manus & caput) componuntur. Ossa durissimæ aridissimæque totius corporis partes existunt. Cartilagines uerò ossibus molliores sunt, secundum ossa tamen durissimæ totius corporis partes censentur, atque his simul et illis uniuersum corpus sustinetur, omniaque ipsis adnascuntur adstabiliunturque. Caluaria cerebri ac sensus organorum sedes, multis ossa constat ossibus: in fronte siquidem sæpius unum tantum, idque præcipuè in mulieribus, conspicitur: in occipitio similiter unum, in uertice duo, ad utramque aurem uel tempus unum, cui cauitas insculpitur, organo auditus reponendo adaptata: & duo complectens ossicula, quorum unum in cudi aut molari denti, alterum uerò malleolo non ineptè assimilaue ris. Hoc quoque os, præterquàm quòd ipsi cartilago auris corpus erectum seruans continuatur, tres exigit processus: ac unus uberis papillam refert: alius stylum, aut acum, aut galli calcar: tertius antrorsum ductus, superiorisque maxillæ ossi in externa oculi sede locato serrata compage commissus, in altero latere portionem constituit eius sedis caluariæ, quam iugale os uocamus. Quinetiam temporis os posteriori inferiorique ipsius parte simul cum occipitis ossis basi eam caluariæ regionem efformat, quam præruptæ rupi non duritia modò, uerùm & imagine conferimus. In capitis basi os insigne habetur, cuneo à Dissectionum professoribus comparatum, & forma impensè uarium, processusque educens uespertilionum alis similimos. Ad hæc, aliud os in narium summo reponitur, cribri aut potiùs spongiæ forma peruium, & cum septem enumeratis ossibus amplitudinem cerebro excipiendo paratam constituens, nariumque septum efformans. In superiori maxilla, præter dentes duodecim ossa recensentur, unum utrinque in exteriori sedis oculi regione, & duo utrinque etiam in interiori, & unum utrobique in inferiori, quod maxillæ ossium longè maximum uisitur, superiorum alterius lateris dentium alueolos sibi insculptos obtinens. Ad palati finem, quà narium foramina in fauces pertinet, utrobique etiam unum eius maxillæ os consistit. & prominens denique nasi pars duobus extruitur ossibus, priuato nomine (ut & cetera maxillæ superioris ossa) destitutis, & nasi cartilagines producentibus, quibus cartilagineæ narium alæ stabiliuntur. Hactenus commemoratorum ossium pleraque inuicem suturis committuntur, ex quibus transuersim in sincipite duæ coronalis dicitur: transuersim uerò in occipitio prorepens, à Λ similitudine nomen obtinuit. quæ autem ab huius uertice ad coronalem per capitis longitudinem incedit, sagittalis uocatur. at quæ ab hac æquè distant, & super aures feruntur, non suturarum speciem, sed ueluti duarum inuicem incumbentium squammarum imagine exprimunt, inde etiam squammosæ conglutinationes appellatæ. Reliqui etiam dictorum ossium connexus nullibi adeò exquisitè, ac tres modò dicti, formam suturæ referunt: imò locis aliquot simplicis lineæ speciem commonstrantes, harmoniæ potiùs quàm suturæ nomen merentur. Inferior maxilla simplici constat osse, nisi forte admodum puerulis in summo mento ex binis ossibus ita coalescat, quemadmodum plurima puerulorum ossa multis ossibus compinguntur, quæ aliàs illis, qui ad summum deuenere incrementum, simplicia occurrunt. Articulatur uerò inferior maxilla utrinque ossi, quod ad aurem consistit, peculiari cartilagine hìc præter illam interueniente, quæ ossium capita sinusque ubi inuicem componuntur, crustæ modo oblinit, & articulum promptè uersatilem & ab iniurijs in

continuo ossium affrictu liberum reddit. Vtraque maxilla plurimùm sedecim dētes nanciscitur, quatuor nimirùm incisorios, duos caninos, & decem molares, qui ut forma anteaquam eruantur apparente differunt, sic etiam impari radicum numero suis præsepiolis infiguntur. In faucibus linguæ radici os præponitur v imaginem uertius quàm Λ exprimens, & pluribus contextum ossiculis, ex quibus humiliora suis extremis aspere arteriæ capitis (quod laryngem libentiùs quàm guttur nuncupamus) cartilagini committitur, quæ scutū refert, primaque laryngis cartilago numeratur, ac tangentibus tota occurrit. Secunda enim magna ex parte posteriorem laryngis sedem constituit, & anulo quem Thraces dextro pollici sagittantes imponunt similis, nomine destituitur, hinc quodammodo id innuniens. Tertia, quæ duabus proprijs extruitur partibus, uasorum quibus aquam manus lauaturis affundimus orificio respondet, in la ryngis medio rimam efformans tibiarū linguis, seu fistularum quas ori indimus rimis similem, lingulamque ob id nuncupatam. His cartilaginibus operculum simul cartilagineum & pingue, ligamentosa natura participans insternit, reliquis interim asperæ arteriæ cartilaginibus, quæ ipsius caudice et ramos in pulmonē digestos cōstituunt, c uel C imaginē exprimentibus. Dorsum medullæ dorsali iter potissimùm paras, ac corporis uclui carina in collum ceruicem ue, thoracem, lumbos, sacrum os, & coccyx seu cuculus os diuiditur, triginta quatuor ossibus (quas uertebras uocamus) efformatum. Ac ceruix septenas cōtinet, multis, & in omnibus non pari numero processibus ornatas, ad quarum primam (quæ sola inter reliquas uertebras spina destituitur, & transuersos processus egregiè protuberantes possidet) caput antrorsùm retrorsùmque rectà ducimus. Secundæ uerò uertebræ (cui processus instar canini dentis prominens accidit) beneficio caput circumuertimus. Reliquarū autem opera in latus, sed tantùm secundariò, caput fertur. Thorax duodecim obtinet uertebras, costarum articulationes admittētes. acharum ferè infima suis ascendentibus & descendentibus processibus, quibus uertebræ præter ipsarum corporum connexum inuicem articulantur, à proximis uertebris suscipitur, uti prima ceruicis & suprà & infrà ipsi contermina ossa suscipit: reliquis interim dorsi uertebris illi duodecimæ superioribus suprà quidem susceptis, infrà autem suscipientibus: & qui inferiùs consistunt, infrà susceptis, suprà suscipiētibus. Subsunt uerò illi duodecimæ, quinque lumborū uertebræ, dein sacrū os, plurimùm sex proprijs compactum ossibus, inuicem ualidēnexis. Coccyx uerò os infima sede in cartilaginē desinens, crebrò quatuor exiguis constituitur ossiculis, nulli neruo iter præbentibus, neque ullo donatis foramine, quò dorsalis medulla reponatur. Pectoris os anteriorem thoracis (qui congruā cordi & ipsi subseruientium organorum sedem parat) regionem sibi uendicans, rarissimè septenis, uerùm paucioribus extruitur ossibus, septem costarum articulationem utrinque excipientibus. Quippe ex duodecim costis utrobique positis, & in cartilaginem degenerantibus, septem elatiores suis cartilaginibus pectoris ossi coarctantur, hinc uerarum & legitimarum nomen sortitæ. Quæ autem pectoris os non contingunt, & tanto magis ab illo anterioris abdominis regione deficiunt, quanto humiliorem sedem obtinent, spuriæ uocitantur. Inferior pectoris ossis sedes in cartilaginem cessat, obtuso ensis mucroni (perinde ac totum pectoris os ensi assimilatur) respondens. Superiori autem ossis sedi ubi latissimum & robustissimū est, iugulumque consistit, utrobique una articulatur clauicula, humeri articulum à thoracis latere remouens. Scapula etenim trianguli imaginem quodammodo exprimens, posteriorem thoracis sedem altera ex parte occupat, in ceruicem desinens, cui acetabulum insculpitur, brachij ossis capiti excipiendo conueniens: dein ex ipsius dorso processum educit, qui super humeri articulum pertingens, summus humerus appellatur: ac clauiculæ peculiaris cartilaginis interuentu (ut & in clauiculæ ad pectoris os nexu sit) articulatus, humeri articulum à thorace auxilio clauiculæ diducit: humeri etiam articulum superius perinde muniens, ac interior scapulæ processus anchoræ, & C aut G comparatus. Brachij itaque os, seu humerus, scapulæ articulatur, & huius inferiori sedi uarijs sinibus & tuberibus exornata, rursus duo inibi ossa iunguntur, radius nimirum & ulna, quæ totius membri no-

B mine

mine etiam cubitus appellatur, pariʻque modo in homine ac in quadrupedibus ad brachiŋ os flectitur, & extenditur, ac in superiori ipsius sede, qua posteriorem cubiti articuli regionem constituit, processum obtinet, quem ʻ gibberum plerịq̃ nominant. Atq̃ ita etiã inferior ulnæ pars alium gerit,ʻquià styli forma nomẽ nanciscitur. Cæte
rùm soli propemodum radioʻ brachiale articulatur, ac ueluti propria cartilagine ab ulna dirimitur, octo ossibus inter se forma & ma
gnitudine omnino differẽtibus extructũ: cuius rursus inferiori sedi
quatuorʻpostbrachialis ossa unà cũ primo pollicis esse coarticulẽt.
Is nanq̃, ut & reliqui quatuor digiti, tribus ossibus seriatim, ac uelu
ti in acie positis cõstituitur. Omniaq̃ digitorũ ossa sunt quindecim.
quibus in secundo pollicis internodio duo parua accedunt ʻossicu-
la, sesami semini collata: qualia etiam in primis quatuor digitorũ in-
ternodiŋs (sed multo minora, impẽsius q̃ cartilaginea) interiori sede
habẽtur: & unum in pollicis tertio internodio, unumq̃ ad exterius
latus connexus ossis postbrachialis paruum digitum sustinentis ad
brachiale secãtibus obseruatur. In reliquis uerò digitorum interno-
diŋs rarius unum, nisi in ualde senibus uisitur. Vtriq̃ sacri ossisʻla-
teri grande os coarctatur: quod ubi amplum est, iliãʻrespicit,ʻillũ
os dicitur: ubi uerò profundiori acetabulo femoris caput admittit,
ʻcoxendicis: ubi autem qua alterius lateris pubem os constituit,
insigniterq̃ perforatum cernitur,ʻpubis os appellatur: quamuis uniuersum os subinde coxendicis nomine comprehendamus.ʻFemo-
ris os suprà rotundo capite ceruici prolixæ, & introrsum obliquæ
tæ adnato, coxendicis ossi inarticulatur: infrà uerò duobus capitibus grandioris in tibia ossis sinus subit, sinu quodam in eorum ca-
pitum medio existente, quo tuber dicti tibiæ ossis recipitur, peculia
ribus etiam cartilaginibus in hoc articulo quadrupedum & auium
genu similimo interueniẽtibus. Iuxta femoris ceruicis externam sedem grandis apparet processus, quemʻnatem & magnum rotato-

rem appellamus. intus quoq̃ alium obtinetʻprocessum, sed multo
minorem exteriori, quem hac occasione minorẽ seu interiorem ro-
tatorem uocamus.ʻIn tibia, quemadmodum & in cubito, duo spe-
ctãtur ossa, quorũ internum crassitie exterius longè uincit, totiusq̃
membri nomine comprehenditur,ʻexteriori interim osse, quod femori nõ articulatur, fibula appellato. Ante tibiæ cum femore articu
lumʻos rotundum, & mola patellaʻue dictum, scuti instar præloca-
tur. Cæterùm imæ tibiæ & fibulæ partes externis ipsorum lateribus
prominentes ac excarnes,ʻmalleoliq̃ appellatæ,ʻtalum eadem sede
in homine ac in quadrupedibus repositum, & eundem in illo ac in
his ussum subeuntẽ apprehendunt: cuiʻcalcis os subŋcitur, tibiæ rectitudinem in posteriora longè excedens. Anterior uerò tali sedes in
rotundum caput definens,ʻossis nauicularis sinum subit, cui tria
ʻtarsiʻossa coarticulantur.ʻQuartum enim exteriusq̃ tarsi os cubi
& tesseræ imaginem exprimens, calci iungitur. Porrò quatuor tarsi
ossibus quinq̃ʻpedŋ ossa coarticulantur, & his quinq̃ pedis digiti
sustinẽtur,ʻè quibus pollex duobus tantũ ossibus efformaẽ, reliqui
interim tria internodia sibi uendicãtibus. Quin etiam in pede (quẽ
homo quadrupedibus multo breuiorem obtinet) totidem ac in ma
nu exigua occurrũt ossicula, quæ sesami semini comparamus: quan
quam primo pollicis internodio hic duo longè grandiora quàm in
manu subŋciantur, horumq̃ interius illud (ut, quod occultæ philo-
sophiæ sectatores corruptioni neutiquam obnoxium esse affirmãt,
& tantisper in terra asseruandum nugaciter contendunt, dum id re-
surrectionis tempore femineis modo hominem producat. Ad hæc,
in pede non minus quàm in manuʻungueis uisuntur, quos non in
opportuné substantiæ gratia huc reponimus: quemadmodum inter partes aliquid suffulcientes referimus palpebrarumʻcartilagi-
nes, cilia quo minus flaccidè concidant erigentes.

DE OSSIVM AC CARTILAGINVM LIGAMEN-
tis, & musculis arbitrarŋ motus instrumentis. Caput II.

LIGAMENTVM non minus quàm muscu
lorumʻeneruationes, & organa quibus animalis spiritus à cerebro deducitur,ʻneruiʻno-
mine uulgòdonatu, corpus est similare, ex osse aut cartilagine principium ducens, sensuʻ
propemodũ omnis expers, durum, sed tamẽ
cartilagine mollius, album, & in colligando,
continendo, obtegendo, ac musculis constituendis uarios in hominum fabrica usus præstans.ʻMusculum enim motus ex nostro arbitratu pendentis instrumentum censetur, multis membraneis liga-
mentisʻnaturam possidentibus, & carne intertextis fibris formatu,
ac ut in se contrahatur, mouendamq̃ partem agat, neruorum à cere
bro animalem uim adferenti opem nõ secus quàm sensuũ organa
requirens: similiterq̃ ut reliquæ quas enutriri oportuit partes, uenis ac arterŋs irrigatũ. Cæterùmʻtendo musculi quædã est eneruati-
o, ac ueluti illius fibrarum citra carnis interuentũ coalitus. Quippe
tendo à musculo tunc tirum producitur, quando ille adeò longam
sedem ab exortu (quem ab osse, aut cartilagine, aut membrana qua-
piam sumit) ad insertionem (quam in mouendam partem molitur)
obtinet, ut ipsius fibræ carnis primariæ musculorum partis implexu uniuerso ductu non indigeant. Adeò ut longiores musculi in
tendinem modò teretem, modò latum, & ueluti membranceum, mo
dò breuiorem, modò longiorem desinant: breues autem musculi
uniuerso progressu carnei permaneant, in nullam notatu dignam
eneruationeʻdegenerantes. Porrò frontis cutem membrana ipsi
substrata, & carneis fibris adaucta mouet.ʻCutis namq̃ natiuum
corporis tegumentum exterius tenuem admodum obnatam exigit
cuticulam, ueluti ipsius uerẽ cutis efflorescentiam. Interius uerò per
uniuersum corpusʻmembrana quædam cuti subducitur, quæ q̃ uã quod
alicubi (quemadmodum hic in fronte) carneis fibris adaugeatur,
carneam uocari uisum fuit. atque inter hanc membranam & cutem
ʻadeps plurimus hominiasseruatur. Superior palpebra etiam dictæ nuper membranæ beneficio mouetur: ea enim parte qua heʻcẽ ad
internam oculi sedis regionem carnea efficitur, palpebram attollit:
qua uerò ad externam oculi sedem carneis fibris ac ʻnostriʻimagi-
nem ductis enutritur, palpebram modò deorsum autor euadit.
Oculum septem agunt musculi, primus in latus introrsum, secũdus
extrorsum, tertius sursum, quartus deorsum ducunt, quintus & sex
tus nonnihil oculum circumuertunt, primi & secundi musculorum

functionem parumper adiuuantes. Atq̃ hi sex forma, quæ oblonga
& quasi teres uisitur, inuicem correspondent, & à dura membrana
neruum uisiorum inuestiente pronati, in anterioremʻdura oculi tunicæ sedem iuxta iridis latera orbiculatim membraneis tendinibus
inseruntur. Septimus uerò his occultus, & solus neruum uisiorum
in orbem ita ambiens, ut sex dicti simul eundẽ cum septimo mu
sculo induebant, in posteriorem duræ oculi tunicæ sedem carneus
implantatur, & ŋsdem ferè motibus cũ sex illis musculis præẽst. Al-
terã nasi alam sursum extrorsumq̃ ducit musculus, ab interiori male
latere pronatus, unaq̃ parte in alam, altera in superius labrum, quã
id alæ subŋcitur, insertus. Introrsum uerò alam contrahit membra-
neus musculus, in narŋ amplitudine sub tunica illas succingẽte latitans. Buccas & labra agunt musculi utrinq̃ quatuor. ac primus
à carnea constituitur membrana, carneis fibris in anteriori potissimùm colli sede & facie ad malas usq̃ enutrita, musculosaq̃ facta.
Secundus à malis pronatus, superiori labro implantatur. Tertius
ab inferiori maxilla in humilius labrum pertinet: Quartus à mentõ
uarius in ea buccarum parte consistit, quam inflamus: atq̃ his accedit musculi illius portio, quem nasi alam extrorsum agere nouimus,
qui unà cum dictis mirificos illos uariosq̃ buccarum & labrorum
motus effingit. Maxillam inferiorem mouentium utrinq̃ etiam
quatuor habentur. primusʻtemporalis est, à uerticis osse, & frontis
quoq̃, & cuneũ referente, & tẽporis ossibus amplo latoq̃ principio
enatus, ac maxillæ inferioris acuto processui insertus. Secundus à
mandendoʻmansorius dictus, ab ea capitis parte que iugale os hinc
cupatur, procedit, externæ maxillæ sedi insertus. Tertius à capitis
processibus alarum imaginem exprimentibus, in internam maxillæ
sedem implantatur, unà cum dictis maxillam attollens, & in latera
prorsumq̃ ac retrorsum agens. Quartus cum suo coniuge maxillã
deorsum trahit, & à stylum imitante capitis processu initiũ ducens,
duobusq̃ uentribus priuatim donatus, ad summum mentum ma
xillæ inseritur. Osʻreferens deorsum rectà trahunt duo musculi
li inuice proximi, atq̃ à pectoris ossis summo pronati. Sursum uerò
id ducit alŋ duo, à maxilla inferiori principiũ ducẽtes: in latera autẽ
id os sursum allicит uno utrinq̃ musculo, qui à capitis processu styli
referte nascit. Deorsum quoq̃ ad latera utrobiq̃ unus mouet musculus, ab elatiori scapulæ costa principiũ obtinens. Octo præsentes
musculi in anteriore ossisʻimagine exprimẽtis sede implantantur.
Linguæ radici à dicti iam ossis medio carnea moles implantatur,
quæ

quæ duorum musculorum uice numerari posset, introrsum'que & deorsum rectà lingua trahit. Ab eius offis lateribus utrobiæ unus etiam linguæ radici inseritur, introrsum, sed tamen magis a latus lin guæ agens. Quintus & sextus linguæ musculi in utroæ latere singu li à capitis proceslibus stylum referentibus enati, linguæ radici im plantantur, eam sursum, prout hic aut ille contrahitur, ad latus du centes. Septimus & octauus utrinæ singuli à lateribus maxillæ infe rioris iuxta molarium dentium radicem enati, linguæ longitudini secundùm humiliora inseruntur, linguæ partem ante sectionem in hiante ore conspicuam deorsum in latera mouentes. Nonus ab in terna maxillæ inferioris sede iuxta summum mentum prodiens, cras sus cernitur, & aliquot donatus inscriptionibus, linguæ'æ humilio ri sedi insertus, eam extrorsum agit. quanquam præter hos muscu los linguæ corpus ante sectionè apparens, talibus implicetur fibris, ut magna Naturæ industria in omnem motus differentiam prom ptilímè feratur. Primam laryngis cartilaginem secundæ commit tunt quatuor musculi, laryngis rimulam arctantes: & quatuor ter tiam cartilaginem secundæ nectunt, rimulam aperientes: & duo ter tiam primæ colligant, rimulam claudentes: alij duo in tertiæ cartila ginis basi consistentes rimulam stringunt. Atæ hi duodecim muscu li, laryngis proprij uocantur. Communiù uerò duo ab osse v simili primæ cartilagini inseruntur, cam'æ antrorsum attollentes, rimulam laryngi referant: & duo à pectoris osse pronati, in eandem quoæ que cartilaginem pertinent. Dein duo à posteriori sede stomachi inuicè proximi principium ducètes, eius quoæ que cartilaginis lateribus to ti carneiæ (ut omnes ferè laryngis) inseruntur, unà cum duobus po stremò dictis laryngem arctantes. Alij duo ab osse v referente ena ti, & in laryngis operculi radicem inseri, id sursum antrorsumæ tol lunt. In caput mouentium classem illi quoæ que reijciantur, qui pri mam ceruicis uertebram priuatim mouent. sunt autem uniuersi sep tem paria, quod et utrobiæ totidem enumerentur musculi. Ad primum par ex quinque superiorum thoracis uertebrarum spinis pronatum, sensímæ extrorsum obliquè ascendens, occipitis ossi im plantatur. Secundum par, quod non duobus, sed pluribus effor mari uidetur musculis, admodumæ uarium cernitur, præcipua ip sius portione ex transuersis quatuor superiorum thoracis, & quin que humiliorum ceruicis uertebrarum procesibus enatum, intror sum'que noñ nihil obliquè conscendens, occipitis ossi implantatur. Tertium à spina secundæ ceruicis uertebræ enatum, & extrorsum parumper obliquè repens, occipitis etiam ossi inseritur. Quartum par similiter occipitis ossi insertum, à prima uertebra inibi prodit, quà aliæ uertebræ in spinam desinunt. Quintum ab occipitis ossis medio transuersim quodammodo ad laterales primæ uertebræ pro cessus fertur. Sextum à spina secundæ uertebræ in eosdem proces sus pertinet, similiter ac quinque nunc enumerata in posteriori cer uicis sede consistens, ac pariter ut tertium, quartum & quintum, mu sculis constit is prorsus carneis & teretibus & gracilibus. Septimum par insignius est, atæ à pectoris ossis summo, clauiculisæ quà illi ar ticulantur, utrinæ unus pronascitur musculus, qui sursum obliquè ductus in mamillarem capitis processum insertionem tentat. Cæte rùm primis quatuor paribus simul tensis caput rectà retrorsum du citur: si uerò trium primorum parium musculi ex altero latere tan tùm egerint, ad capitis circumuersionè auxiliabuntur, & in illo mo tu quintum & sextum par, primam ceruicis uertebram unà cum ca pite in gyrum ducent. Septimo uerò paris musculis suum munus fir m il obeuntibus, caput rectà antrorsum flectitur: qui autem alter natim laborant, circumuersionis autores efficiuntur. Verùm ad ca pitis motum, quo id secundariò cum ceruice flectitur, & extenditur, ac in latus ad humeros ducitur, ceruicis musculi famulantur, inter o cto paria dorsum mouentia reponendi. Ac primum par à quintæ thoracis uertebræ corporis lateribus incipiens, ad primam usæ cer uicis uertebram pertingit, stomacho subiectù, superioremæ dorsi partem flectens. Secundum à prima thoracis costa utrinque prona tum, & internæ transuersorum ceruicis uertebrarum procesuum sedi insertum, ceruicem in latus, sed antrorsum magis, ducit. Tertiù ex transuersis sex superiorum thoracis uertebrarum procesibus o riginem obtinens, externæ sedi transuersorum uertebrarum ceruicis procesuum implantatur, ceruicemæ que retrorsum inclinantem ad latus agit. Quartum par à septimæ thoracis uertebræ spina ad secundam usæ ceruicis uertebram pertinet, omnibus intermedijs uertebris, ut & primum par, insertum, ac ex eisdem principium ca piens, superioremæ que dorsi partem extendens. Quintum par u trinque etiam habet musculum, ab ilium osse in transuersos lumbo rum uertebrarum processus, & infimam thoracis costam insertum, inferiorisæ dorsi partis flexus opificem. Sextum ab infima offis sa

cri sede in dorso incipiens, ad ceruicem usque fertur, in transuersos processus lumborum, sed manifestiùs adhuc thoracis uertebrarum insertum. Huius paris ambobus musculis contractis, dorsum exten ditur: altero autem tantù laborante (ut & in alijs paribus) hoc quo que par obliqui seu in latera motus autor euadit. Septimum sub sex to occultatum à posteriori sacri offis sede principium ducens, ad un decimæ usæ que thoracis uertebræ spinam conscendit, omnibus in termedijs spinis asserum, ac spinas inuicem colligendo, dorsum hic extendens: uti & in sua sede octauum par, quod ab undecima thora cis uertebra ad septimam usæ ceruicis porrectum, ita prorsus in termedijs spinis committitur, ut septimum illis quibus adnascitur spinis. Scapulam adpectus mouet musculus à thoracis secunda, tertia, quarta & quinta costis anteaquam in cartilaginem illæ finiit enatus, & trianguli modo in internum scapulæ processum insertus. Secundus scapulam mouentium ab occipitio pronatus, & dein se cundùm ceruicis longitudinem ad octauæ usæ que thoracis uertebræ spinam à uertebrarum spinarum apicibus principium sumens, in scapulæ spinam summumæ que humerum & clauiculæ portionem inseritur, tota sui parte que in ceruice consistit scapulam sursum uel lens: ea uerò quæ sub ceruice in posteriori thoracis sede habetur, monachorum que cucullis respondet, scapulam deorsum trahens. Tertius à transuersis superiorum ceruicis uertebrarum procesibus enatus, ac elatiori scapulæ basis angulo insertus, illam quoæ que attollit. Quartus ex quintæ, sextæ & septimæ ceruicis, ac triù primarù tho racis uertebrarum spinis præcipuè enascitur, scapulæ'æ basi inser tus, eam ad dorsum nonnihil sursum contrahit. Primus brachij motuum autor à media clauiculæ sede pectoris ossi proxima, & pe ctoris osse pronatus, ac quodammodo in angulum arctatus, bra chium pectori adducit. Secundus ab altera clauiculæ parte, & sum mo humero, & scapulæ spina principium sumens, suoæ que uertice bra chij ossi transuersim insertus, ad brachium uarie attollit, hunc humeri articulum pulchrè 'tegens, & figuræ △ non ° absimilis. Tertius ab humiliori scapulæ costa prodiens, brachium rectà ad dorsum alli cit. Quartus à sextæ thoracis uertebræ spina ad humiliorem usæ que sacri offis sedem ex intermediarum spinarum apicibus originem du cens, & uelutiin triangulum uerticem coactus, in brachium inibi inse ritur, ubi tres commemorati in insertionem longè sub offis capite sca pulæ articulato moliuntur. atque hic musculus brachium uarie de orsum trahit, utiipsius principum perquam amplum est: neque ex puncto ita ducitur, ut musculi simplicem prorsus motum obituri. Quintus uniuersam scapulæ cauum costas respicies occupat. Sex tas gibbam scapulæ sedem uniuersam sibi uendicat, sub ipsius spi na consistentem. Septimus sinum implet inter scapulæ spinam & elatiorem ipsius costam conspicuum. Hi tres amplis implantationi bus in ligamenta humeri articulum ambientia inseruntur, brachij circumactus opifices, quamuis septimus & ad brachij eleuationem aliquod auxiliari uideatur. Thoracem mouentium primus à clauicula enatus, in primam thoracis costam inseritur, illam sursum ducens, hacæ que occasione ad thoracis dilatationem iuuans. Secun dus à basi scapulæ pronatus, tanquam digitis quibusdam octo su perioribus costis inuicem adnexam in cartilaginem degenerant, in seritur, eas'que extrorsum mouens, thoracem dilatat. Tertius ex a picibus spinarum trium inferiorù ceruicis, & primæ thoracis uerte brarum latum membranæ'que principium ducens, ternis ue luti digitis, tribus quatuor superiorum costarum interuallis sub scapulæ basi inseritur, eas'que costas sursum obliquè ducens, thora cem ampliat. Quartus ab ilium osse incipiens, sursum'que ad cer uicem ductus, duodecim costis quà primum à uertebris discedunt, inseritur, thoracem arctans. Quintus ex spinarum apicibus dua rum infimarum thoracis & aliquot lumborum uertebrarù mem branæus enascitur, & transuersim ductus, ac nonæ, decimæ & un decimæ costis ubi in interiora reflectuntur insertus, thoracem di latat. Sextus in thoracis amplitudine repositus, uerarum costarum offium interuallis consistunt, à superiori costa antrorsum obliquè suas fibras in humiliorem mittunt: interiores uerò ab inferiori co sta sursum obliquè in anteriora, ad superiorem costam fibras porri gunt. In sex uerò cartilaginum quæ legitimis costis affcribuntur interuallis, exteriorum musculorum fibræ ab humiliori cartilagine in superiorem obliquè antrorsum repunt: interiores autem fibræ à superiori cartilagine ad inferiorem antrorsum pertinent, ac pro inde in sex legitimarum costarum interuallis quaterni numerantur musculi. in spuriarum autem costarum interuallis bini tantum, uni

C uersiæ

uerſiſ́ꝗ unius lateris intercoſtales muſculi quatuor ſunt & triginta, ad unum omnes arctando thoraci præfecti. Atque numeratis hacte nus in uno latere thoracis quadraginta muſculis, alij totidem in altero latere congruunt. & his octuaginta unus accedit, utrique lateri communis, ipſum uidelicet [m] ſeptum tranſuerſum infimæ pectoris oſſis & ſpuriarum coſtarum cartilaginibus et uertebris lumborum ſuperioribus inſertum, ac in medio nerueum, in circuitu uerò ad in ſertionem carneum, ut ſanguificationi generationi que ſubſeruien tia organa à cordis & partium ipſi ſubminiſtrantium ſede dirimẽs, thoracemꝗue proprio munere dilatans. His accedunt octo ab dominis muſculi, utrinque nimirum quatuor. Primus ſeu extimus obliquè deorſum in priora fibras porrigit, cum ſuo coniuge toti ab domini inuolucrum efformans. Secundus obliquè ſurſum in anteriora fibras mittit, & cum ſuo pari inuolucrum quoque abdomini conſtituit. Tertius recta fibras ſurſum ducens, à pubis oſſe ad pectus conſcendit. Quartus tranſuerſim fibras digerit, ac cum ſuo pari etiam, ut & obliqui, inuolucrum abdomini extruit, ad thoracis conſtrictionem non minus quàm cæteri abdominis muſculi ſuppe tias ferens. Cubitum duo flectunt, quorum anterior unum ca put à ſcapulæ ceruicis elatiori ſede, alterum ab interno ſcapulæ proceſſu mutuatur, & illis efformatus capitibus in radium inſeritur. Poſterior à brachij oſſe enatus, in anteriorem cubiti articuli ſedem, potiſſimùm uerò in ulnam inſeritur. Extendunt autem cubitum tres, ac unus ab humiliori ſcapulæ coſta naſcitur, ſecundus à brachij oſſis ceruicis poſteriori ſede. hi in deſcenſu inuicem connaſcun tur, & ipſis tertius admiſcetur, à media ferè longitudine oſſis brachij enatus, & ſimul cum illis in poſteriorem ulnæ proceſſum inſer tus. In interna cubiti ſede gracilis reponitur muſculus, qui ab in terno oſſis brachij tubere pronatus, in latum degenerat tendinem, interneꝗue ſummæ manus cuti magna ex parte ſubnatù: cuius beneficio ea cutis minus uerſatilis, & ad tangendù aptior reddi credit̃. Radius in pronù duobus muſculis ducitur. uno quidẽ ab interna cubi ti articuli ſede enato, & in radiù obliquè implãtato: altero autẽ iuxta brachiale ab ulna in radiù tranſuerſim ducto. In ſupinù uerò radius alijs duobus fer tur, uno quidem longo, & à brachij oſſe ad inferio rem radij partem, cui brachiale articulatur, pertinente: altero ab externa cubiti articuli regione ad mediam radij longitudinem obli què exporrecto, ibidem que inſerto. Brachiale quatuor præcipuis agitur muſculis, ac duo primi ab interno tubere brachij oſſis pronaſcuntur, & unus poſtbrachialis oſſi indicem ſuſtinenti, alter uerò brachialis minimo oſſi implantatur. tertius à brachij oſſe ena tus, bifido tendine in poſtbrachialis oſſa indicem & medium ſuſti nentia inſeritur. quartus ab externo brachij oſſis tubere pronatus, ac ulnæ exporrectus, in poſtbrachialis os paruù ſuſtinens digitum inſertionem molitur. Duo primi ſimul brachiale flectunt: ter tius uerò & quartus ſimul contracti, id extendunt. Verùm primo ſimul cum tertio tenſo, brachiale in internum latus agitur: ſecundo autem & quarto unà laborantibus, id in externum latus inclinatur. Digitos manus mouentium primus ab interna anteriori ꝗ cubiti articuli ſede enatus, ante brachialis radicẽ in quatuor diſſcinditur tendines, ſecundis quatuor digitorum internodijs inſertos, illaꝗ́ fle ctentes. Secundus ab eadem ſede cù primo, ſed magis decliuè, origi nem ducens, primo ſubſternitur, in quatuor etiam diremptus tendi nes, qui primi tendinibus ſubiecti, ante ſecundi digitorù internodij radicem eos tendines perforãt, tandemꝗ in tertia quatuor digitorù oſſa inſertionem molientes, ea flectunt. Tertius à radio iuxta cubiti articulum enatus, tertio pollicis articulo inſeritur, ipſius que flexus autor euadit. Tertium muſculum, in ſerie alij tredecim numero ſub ſequuntur, in extrema manu repoſiti, quorum bini primo ſingulo rum quinque digitorum oſſi inſeruntur, eius flexionis opifices, & tres ſecundo pollicis internodio potiſſimùm inſerti id quoque fle ctunt. Decimusſeptimus itaque digitos mouentium ab externo brachij oſſis tubere natus, & indici, medio ac anulari præcipuè im plantatus, eos digitos extendit. Decimusoctauus ab eadem qua nu per dictus ſede prodiens, parui digiti extenſionis primarius autor eſt, & uariè cum decimaſeptimi tendine tribus anularis digiti oſſi bus inſerto commixtus, etiam abductioni uerſus exteriora nonni hil inibi ſubſeruit. Decimusnonus cum illo qui uigeſimusprimus erit, commune principium ab ulna, non procul à brachialis ſede nanciſcitur: & in duos ferè diſſectus tendines, unum exteriori indi cis, alterum medij lateri digerit, illorum que digitorum in exter num latus abductionis autor cenſetur. Vigeſimus à brachiali pro natus, & externo poſtbrachialis oſſis paruum ſuſtinentis lateri ex porrectus, primo que eius digiti oſſi inſertus, illum extrorſum in latus abducit. Vigeſimusprimus externo lateri exterioris pollicis

regionis ad tertium uſque articulum implantatur, pollicis uerſus in dicem extenſionis autor. Vigeſimusſecundus ab ulna paulo ſupe riùs quàm nuper dictus prodit: & mox bipartitò ſciſſus, una por tione in tendinem ceſſat, brachialis oſſi pollicem ſuſtinenti inſer tum, & ductum, quo manus radij motum in pronum ſequitur, ad iuuantem. Altera autem pars in duas rurſus dirimitur, quæ ſin gulæ unum efficiunt tendinem: & hæc quidem ſuum tendinem primi pollicis oſſis exterioris ſedis interno lateri inſerit: illius uerò partis tendo illi oſſi tantum obnaſcitur, in ſecundum & tertium pollicis os inſertus. Vigeſimustertius ad internù primi pollicis oſſis latus ſedem obtinens, pollicè ab indice inſigniter abducit. Vigeſi musquartus ab oſſe poſtbrachialis indicem ſuffulciente enatus, & primo pollicis oſſi præcipuè inſertus, pollicem indici proximè ad ducit. Superſunt adhuc quatuor graciles muſculi, quatuor tendi nibus ſecundi digitos mouentium muſculi in uola attenſi, & in internum latus primi quatuor digitorum oſſis inſerti, eorum que digitorum ad pollicem in latus adductionis miniſtri. Reponun tur itaque in interna cubiti ſede muſculus, latum manus efficiens tendinem, primus & ſecundus brachialis motuum autores, pri mus ſecundus & tertius digitos mouentium, & duo ultimi in pronum ducentes. In externa autem ſede conſiſtunt decimusſep timus, decimusoctauus, decimusnonus, uigeſimusprimus, uigeſi musſecundus digitos mouentium, & tertius & quartus brachia lis motibus præfectorum, & duo radium in ſupinum agentes, ſunt que omnes numero nouem. Verùm decem conſurgent, ſi à uigeſimo ſecundo digitos mouentium illam portionem diſtinxe ris, quæ brachialis oſſi pollicem ſuſtinenti tendinem offert. In extre ma manu obſeruantur decem muſculi, primos articulos digitorum flectentes, & tres ſecundi pollicis internodij flexus autores, & dein digitos agentium uigeſimus, uigeſimustertius, uigeſimusquartus, & quatuor muſculi quorum beneficio quatuor digiti pollici addu cuntur. Singuli uirorum teſtes cum ſuis ſeminalibus uaſis tu nica obducuntur à peritonæo pronata, & aliquot rectis carniſꝗ fibris enutrita, infimæ que ſemen deferentis uaſis ſedi inſerta. His fibris teſti [n] unus conſtituitur muſculus, cuius ope teſtis obſcurè ſurſum uellitur. Sic & membranæ uterum firmantes, utrinque carneis donantur fibris, hac que ratione uterus utrobique unum poſſidet muſculum, cuius auxilio leuiter ſurſum uerſus ilia contra hitur. Veſicæ ceruici unus orbiculatim obnaſcitur muſcu lus, intempeſtiuam urinæ excretionem prohibens. Item alius quo que [o] muſculus, recti inteſtini finem circulatim ambit, immaturæ egeſtioni præfectus, & duo alij muſculi poſt egeſtionem rectum inteſtinum alacriter ſurſum attollunt. Penis radici utrinque à pubis oſſe gracilis inſeritur muſculus, nimis quàm ſegniter ad ip ſius tentiginem iuuans. Quinetiam ab anteriori ſede muſculi re ctum inteſtinum orbiculatim amplectentis, duo enaſcuntur muſcu li, inuicem ſibi proximi, & urinæ meatui, quà is ſub pubis oſſibus ſurſum reflectitur, implantati, meatum que in ſeminis eiaculatione quo minus ille à flexum occludatur, dilatantes. Primus femori mouentium ab ilium oſſis extima ſede, & coccygis oſſis poſterio ri regione enatus, poſteriori magni femoris proceſſus ſedi, & eius quoque radici ampla inſertione innaſcitur. Secundus ſub primo magna ex parte reconditus, magiſ́ que ab anteriori ilium oſſis ſede pronatus, etiam magno femoris proceſſui inſeritur. Tertius ſecun do longè minor, ipſoꝗ penitus occultatus, ab ilium oſſe iuxta po ſteriorem coxendicis acetabuli ſedem enaſcitur, magno etiam femo ris proceſſui inſertus, & ut duo priores femur extendens, extror ſumꝗ in latus agens. Quartus à tribus humilioribus oſſis ſacri oſſi bus pronatus, etiam magno proceſſui inſeritur, femur extendens, & extrorſum non parum circumducens. Quintus omnium cor poris muſculorum maximus, & pluribus partibus à coxendicis oſſe ac pubis principium ducens, poſteriori femoris ſedi & hu miliora ipſius uſque capita inſertus, extenſionis & rectæ ſtationis autor habetur, femur etiam introrſum agens, idꝗ ipſius potiſſimù portione à pubis oſſis humiliori ſede pronatus. Sextus à duabus in fimis thoracis & aliquot ſuperioribus lumborù uertebris initium capiẽs, minori femoris proceſſui inſeritur, femoris flexus unà cum ſeptimo autor: qui à tota interiori oſſis ilium ſede enatus, minori quoque proceſſui elatius ſexto implantatur. Octauus à pubis oſſe pronatus, ſub minori proceſſu femori longa inſertione implanta tur, id quidem flectens, uerùm etiam impenſe introrſum mouens. Nonus anteriorem pubis oſſis foraminis ſedem occupans, maio ri que femoris proceſſui inſertus, introrſum femur circummuertit. Decimus poſteriorem interioremꞌue dicti iam foraminis ſedem fi

bi

m Διάφραγμα
ꝗ ϕϱυϵϲ. ϗ
ἀϖϖ λαϲϑια.

n λίϗϲϱε
μαεύϲϱε.

ο ϴϱϱϖϊϱϵϲ

bi uendicans, & secundùm coxendicis ossis posteriora tutissimè
reflexus, ac uelutialijs musculis hîc pronascentibus adauctus, maiori'que femoris processui implantatus, femur extrorsum circum-
uertit.   Tibiam agentium primus ab ilium ossis spinæ antica sede
enatus, & admodum obliquè secundùm femoris internam sedem
ρ ἀντι-
κνημιον. ducitur, anteriori tibiæ sedi inferitur, totius'q corporis gracilimus,
& simul longissimus musculus efficitur. Secundus à pubis ossium
commissura enatus, in eandem cum primo sedem inseritur. Tertius à coxendicis ossis appendice principium obtinens, in eam quoque sedem tibiæ implantatur. Quartus ab eadem sede coxendicis
ossis pronatus, & portionem sui à femoris osse in descensu assumes,
tibiæ cum fibula articulo, sed potissimùm fibulæ inseritur. Quintus
ab eadem etiam regione principium adeptus, in anteriorem tibiæ sedem cum tribus primis, uerùm minus decliuis, implantatur. Sextus
ab ilium ossis spina pronatus, membranæ modo musculis os femoris amplex ntibus obducitur, & potissimũ ad exterius latus genu
articuli inseritur. Septimus à radice magni femoris processus enatus, & externum femoris latus occupãs, cum octauo & nono tendi
nem efformat, cui patella innascitur. Octauus enim à ceruice femoris & radice maioris ipsius processus originem duces, quasi totum
femoris os proximè ambit. Nonus à coxendicis ossis tubere supra
ipsius articulum, anteriori in parte cõspicuo originem
sumens, & septimo & octauo instratus musculis, ad anteriorem genu regionem fertur : ac in tendinem degenerans, cum duobus nuper dictis unum constituit tendinem, anteriori tibiæ sedi ualidissimè
implantatum. atque ita primus, sextus, septimus, octauus & nonus
tibiæ extensionis opifices habentur, secundo interim, tertio, quarto
& quinto tibiam liquidè extendentibus .   Qui in poplite latitat
musculus, & ab externo genu articuli ligamento in tibiæ os oblique
pertinet, tibiam non flectit, & si quid agit, obscurè motum imitatur
primi radium in pronum mouentis musculi.   Eorum qui pedem
mouent, primus ab interno femoris capite iuxta genu articulum incipit, uti & secundus ab externo capite principium nanciscitur, ambo posteriorem suræ partem constituentes, & cum quarti pedem
mouentis musculi tendine congredientes, in calcem inseruntur. Ter
tius exiguus est musculus, ab externo quoque femoris capite pronatus, & in poplitis regione adhuc in tenuissimum tendinem desinens, calcis ossis interno lateri insertum . Quartus, pedem mouentium maximus est, ac à fibulæ cum tibia coarticulatione incipiens,
in ualidissimum desinit tendinem, cui duorum primorum tendo
unitur, simul cum ipso calci insertus. Quintus musculus tibiæ ac fibulæ posteriori sedi proximè allocatus, ab ijsdem pronascitur ossibus, ubi illa primùm dehiscunt, & iuxta interioris malleoli posteriora tendinem educit in tarsi os, quod tesseram imitãti ossi conterminum est, insertum. Sextus in anteriori tibiæ sede positus, à tibiæ osse
quà illi fibula superiùs coarticulatur pronatus, tẽdinem in radicem
pedṃ ossis pollicem sustinentis inserit . Septimus à fibula principiũ
ducens, externum tibiæ latus occupat, ac tendinem sub pedis inferio
ra reflexum, pedij ossi pollicem sustulcienti inserit. Octauus septimo occultior, atque à fibula etiam enatus, tendinẽ inserit radici ossis
pedṃ paruum digitum sustinentis. Nonus pars est eius quem quatuor pedis digitos extendere mox dicturus sum : inseritur uerò ad
m dium ferè longitudinis ossis pedij, antè paruum digitum positi.
Primis quinque musculis pes extenditur, seu fortiter terræ admoue
tur, quamuis tertius musculus id inualidè præstet : & si quid ad pedis motum facit, cum in obliquum uersus interiora agit: sexto autẽ,
septimo, octauo & nono pes flectitur, horumq musculorum beneficio pes laterales motus, prout singuli agunt musculi, experitur.
Pedis digitos mouentium primus in planta totus collocatur, & in-
tima parte membranam impensè crassam, & lato in manu tendini
non absimilem pertinaciter obnatam possidet : & à calcis osse pronatus, secundis quatuor digitorum internodijs singulos tendines
offert, eorum flexus autores. Secundus & tertius posteriorem tibiæ
sedem perreptant, & secundus tibiæ ossi magis quàm fibulæ exporrectus, ab illoq pronatus, tendinem in plantam mittit, qui in quatuor dissectus tendines, singulos quatuor digitorum tertijs ossibus
inserit, primi musculi tendines, perinde ac in manu sit, perforantes,
eaq ossa flectentes. Tertius à fibulæ ad tibiam nexu enatus, & etiã
fibulæ multum exporrectus, tendinem in plantam profert : à quo
exigua portio tendinibus commiscetur, indicis & medij tertium internodium flectentibus: reliquus autem totus in secundum pollicis
os inseritur, illius flexionis autor. Decem musculi his succedunt,
qui insigniter mutuò implexi, ac pedij ossibus adnati, prima digitorum ossa flectunt, duobus musculis ad singulos digitos pertineti-
bus. Decimusquartus igitur, cuius pars nonus pedem mouentium

censebatur, à tibiæ anteriori sede pronatus, in quatuor dirimitur
tendines, qui quatuor digitis inserti, eorum extensionis autores cen
sentur. Decimusquintus ab anteriori tibiæ sede etiã procedens, pollici inseritur, ipsius extensionis opifex . Decimussextus in superiori
pedis sede reponitur, carneaq est moles in quatuor diuisa tendines,
quorum unus pollicis superioris sedis externo lateri implantatur, se
cundus indicis, tertius medij, quartus anularis : atque hi tendines
eorum digitorum in externum latus abductionis opifices sunt. Decimusseptimus externum pedis latus occupans, & parui digiti primo ossi insertus, illum à cæteris remouet. Decimusoctauus interno
pedis lateri exporrectus, pollicem à cæteris digitis abducit. Dein in
pedis planta carnea consistit substantia, in quatuor graciles dissecta
portiones tendinibus adhærentes, quorum ministerio tertia quatuor digitorum ossa flectuntur. Hæ portiones interno quatuor digitorum lateri ad primum articulum insertæ, eorum ad pollicem ad
ductionis autores habentur. ac proinde si istas quatuor portiones
quatuor musculorum uice enumeraueris, in tibiæ posteriori sede
obseruabis primum, secundum, tertium, quartum pedem mouentium, secundum & tertium digitos mouentium, et sub illis quintum
pedem mouentium. In anteriori uerò sextum, septimum, octauum,
nonum pedis motuum autores, & decimumquartum, decimum-
quintum digitos mouentium: in pede autem habueris primum digitos agentem, & decem prima digitorum ossa flectentes, & decimumsextum, decimumseptimum, decimumoctauum digitos quoque mouentium , nisi decimumsextum in plures diuidere propositũ
esset.   Porrò in musculorum enarratione passim ligamentorum
non memini , quòd magna ex parte articuli inuicem respondeant.
Omnibus enim articulis orbiculatim ligamentum ab uno osse in
aliud, aut in cartilagine, uel à cartilagine in os, aut cartilaginem in
sertum priuatim obducitur, paucis que articulis ligamenti accesse-
re peculiaria. Vt in capitis articulo teres quoddam à dente secundæ
ceruicis uertebræ in occipitis os ducitur, & secundũ dentis posteriorem sedem in prima uertebra unum transuersim fertur . Verte-
brarum corpora ligamentis admodum cartilagineis committuntur, ascendentes uerò & descendentes eorundem processus ualidis
quoque, sed tantùm ambientibus colligantur : dein in spinarum interuallis membraneum cõsistit ligamentum, sicut in cubito & tibia,
ubi ossa hîc inuicem dehiscunt : insuper in pubis ossium foraminibus huius generis occurrit ligamentum, seu membrana potius . In
humeri articulo tria peculiaria uisuntur, quorum primum teres est,
& ab interno scapulæ processu enatum , in externum humeri caput
fertur, alia duo ab elatiori ceruicis scapulæ sede enata, in idem caput
pertinẽt, & unũ quoq hîc ab inte no scapulæ processu in summum
humerũ ducitur. In brachiali ossium inter se & cum postbrachialis
ossibus connexu, ut & in pede, passim cartilaginea ligamenta inter-
ueniunt. A' sacro osse duo teretia in coxendicis os pertinent. Ex femoris superiori capite teres ligamentum in coxendicis acetabulum
inferitur . In genu articuli medio cartilagineum consistit ligamen
tum, & dein in posteriori ipsius sede, & utrinque ad latera unum
peculiare secantibus obuium est. Ex ligamentorum uerò transuersim tendinibus obductorum , tendines que ne sua sede declinent,
continentium numero, in interna brachialis sede unum consistit, &
secundùm uniuersam internam cuiusque digiti sedem unum continuum habetur, iuxta brachialis radicem in externa radij & ulnæ sede sex occurrunt . In anteriori tibiæ sede iuxta talum unum quoque
obseruatur, & tria inter calcem internum'que malleolum , dein u-
num inter calcem & externum malleolum . Sic & in interna inferiori'ue pedis digitorum sede, huius quoque generis ligamenta spectantur.

DE ORGANIS NVTRITIONI,
quæ ex cibo potuq; fit, famulantibus.
Caput III.

VANDOQVIDEM homo propter semen
genitale sanguinemq menstruum, generationis
nostræ primordia, & eorum quibus constamus
materiam immortalis fieri nequiuit, immensus
rerum Opifex, ut is quàm diutissimè uiuat, &
species ipsius nunquam deficiens semper immor
talis perseueret, solicitè machinatur. Vt enim ad
debitum incrementum homo attingeret, ac illa quæ insitus calor
perpetuò depascitur, quàm fieri posset proximè restituerentur, or-

D   gana

gâha nutritioni multifariam famulantia obtinuimus. Cibus namq́ dentibus, quò poſtmodum leuiori opera conficiatur, effractus, per inde ac potus ab ore in uentriculum, tanquam in promptuarium, per ᵃ uiam ducitur, quæ duabus proprijs tunicis tendi, atque in ſe concidere aptis formata, ex faucibus ſub aſpera arteria, ac dein ſecundùm thoracis uertebras per ſeptum tranſuerſum in ſuperius, ſiniſtrum ue uentriculi orificium pertinet, & ſtomachus aut gula uocatur. ᵇ Ventriculus uerò inter iecur & lienem ſub ſepto repoſitus, & inſigniter capax, & ſecundùm tranſuerſum oblōgus, in ſiniſtraq́ corporis ſede quàm in dextra amplior, & duabus demum tunicis diſtendi contrahiq́ue idoneis, & tertio quodam inuolucro quod peritonæum educit intectis, efformatus: & compluribus uenis, arterijs ac neruis implexus, quod ab ore illi delatum eſt concoquit, ac ueluti in lacteum cremorem inſita ui emutat, quem per ſuum ᵈ inferius orificium ex elatiori ſede dextri ipſius ᵈ inteſtina propellit. quæ corpora ſunt teretia uno continuo quæ ac innumeris orbibus gyriſque tortuoſo ductu à uentriculo ad podicem pertinentia, & ſimiliter duabus proprijs tunicis extructa. quibus tertia à peritonæo accedit, non minus quàm duæ propriæ, laxari contrahiq́ue apta, non tamen undique pariter ampla. Inteſtinorum ſiquidem à uentriculo procedens origo, ſecundùm uentriculi poſteriora ad dorſum uſque reflexa, & ᶠ duodenum nobis appellata, ac dein huic ſuccedens inteſtinorum pars, quam ᵍ ieiunum dicimus, & ea quæ ilium ſeu uoluulus nūcupatur, ſuiſq́ orbibus ilia & ſedem undique umbilico ſubiectam & contorminam implet, pari quodammodo conſtat amplitudine, quæ quum arcta ſit, ut dictis nuper inteſtinorum partibus gracilium nomen deretur, in cauſa fuit. Porrò inteſtinorum pars in quam ilei terminus ceſſat, impenſe craſſa & ampla uiſitur, ipſiq́ appendiculum continuatur, inſtar lumbrici inuolutum anguſtumque, & uno ore donatum, ideoq́ ᶦ cæcú diſſectionis proceribus appellatum. Ipſa uerò craſſa inteſtinorū pars à dextri renis ſede ad iecoris cauû aſcendens, hinc ſecundùm uentriculi fundum ad lienis ſedem, & illinc ſecundùm ſiniſtri renis ſedem deuoluitur, & gyro quodam in ſiniſtra pubis ſede reflexa, toto que illo ductu ᵏ colum conſtituens, ſuper ollis ſacri initium recta dorſum ad anum fertur, ˡ recti ac principis inteſtini nomen inibi obtinens. In hæc itaque inteſtina quicquid uentriculus confecit, depellitur, per uarios illorum gyros deuoluendum: uenæ autem quæ innumera ſerie à iecoris cauo unà cum arterijs à magna arteria depromptis, inter duas membranas inteſtina dorſo colligantes, multaq́ue pinguedine ac ᵐ glandulis abundantes, & ᵑ melenterium uocatas, in inteſtina pertingunt, ex ipſis (ſed potiſſimùm gracilibus) inteſtinis, quicquid ſanguini deuiciando idoneum eſt, ſimul cum aqueo tenui q́ue concoctionis uentriculi recremento exugunt, id iecori ſanguificationis officinæ deferentes. Ipſum autem craſſius & ineptum ſuctioni recrementū ſenſim in craſſa colligitur, impenſe craſſa & ampla uiſitur, illic tantiſper aſſeruandum, donec hominem moleſtans, recluſo recti inteſtinū orbiculatim ambiente muſculo, ſemel uniuerſim que hominis arbitratu egeratur. ᵒ Iecur in nullas diremptum fibras lobos ue, ipſi ſubminiſtrantium organorum elatiſſimam ſedem occupat, ac uentriculo magna ex parte incumbens, ſepto que tranſuerſo proximè ſubditum, & magis dextram quàm ſiniſtram corporis ſedem implens, ſuprà gibbum, infrà cauum uiſitur, accumbentium illi partium formæ ad amuſſim congruens, & multarum uenarum implexu, quibus propria iecoris ſubſtātia concreto nuper ſanguini ſimilis circunfunditur, efformatum, tenuiq́ue inuolucro ab ipſius ligamentis quibus peritonæo firmatur procedente obtectum, neruulos que duos & arteriam item unam admittens, & naturalis ſeu altricis, aut, ut Plato dicebat, uenereorum cibo rum & potuum concupiſcibilis animæ ſomes. Cæterùm uenarum per iecur diffuſarum una ſeries in gibbo ipſius conſiſtit, ad ᵖ cauam pertinens uenam, altera in iecoris cauo reponitur, portæ uenæ cau dicem conſtituens: quæ primum ueſiculæ bilem flauam recipientị ſurculos duos promit, dein uentriculi poſteriori ſedi iuxta inferius ipſius orificium: inde uentriculi fundi dextræ parti ramus offertur, à quo ramuli in uentriculo & ſuperiorem ᵠ omenti membranam ſparguntur, quod membraneum eſt corpus ſacculi modo extructum, & uaſis tutò deducendis præcipuè adaptatum: quanquam quum uenis arterijs que & pinguedine illis affuſa ſcateat, inteſtinorum quoque calori fouendo auxilietur. Inſtar circuli enim à dorſi medio ſub uentriculi poſteriori ſede incipiens, per iecoris cauum ad fundum uentriculi (à cuius tertio inuolucro inibi pronaſcitur) ad liẹnis cauum, & hinc ad dorſi medium ſuo ueluti initio defertur. Hinc uerò ſacculi inſtar deorſum protenſum, anteriori inteſtinorum ſedi obuoluitur, innataẹ ue, colum inteſtinū quà uentriculo exporrigi

tur, meſenterij uice dorſo committens. Cæterùm portæ caudex poſtquã omento ſuffultus, dictas nuper depromplit propagines, in duos partitur truncos, ac dextrum qui grandior eſt, per melenteriū uariè digeſtum, inteſtinis offert, prius duodeno inteſtino, & ieiuni initio ramum exhibens, glanduloſo corpore huic inteſtinorum ſedi exporrecto ſuffultum. Siniſter truncus inferiori membrana omenti intectus, uentriculi poſteriori ſedi quà dextram porta reſpicit, paruam offert propaginem, dein etiam omenti inferiori membranæ, mox ᵍ glandulis hic uaſorum tutæ diſtributioni præfectis, & colore carneis: dein ſecundùm poſteriora uentriculi ramus ab illo aſcendit, uentriculi dorſi medium ſpectãti ſedi ſurculos primùm exporrigens, & coronæ ritu ſuperius uentriculi orificium amplectens, à quo præter ſurculos deorſum ſurſumq́ depromptos, unus ſecundùm uentriculi poſteriora ad inferius uentriculi orificiū prorepit. Ipſe autem ſiniſter portæ caudicis truncus ſiniſtrorſum ſemper tendens, inſignem hic quoque promit uenam, omento & colo inteſtino implicitam: ille uerò in uarias ſectus propagines, & ſobolem adhuc humidiori membranæ ſpargens, lienis cauo inſeritur, ab ipſius propaginibus, anteaquam lienem ſubeant, uentriculi ſiniſtro lateri ramulos offerens: inter quos inſignis occurrit, qui fundum uentriculi in ſiniſtra ſede perreptans, uentriculo & ſuperiori membranæ omenti ſoboles deriuat. Venæ portæ autem ſoboles per iecoris ſubſtantiam diſtributæ, quicquid ab inteſtinis & nonnihil à uentriculo quoque iecori detulerunt, in ſe continent: ac iecur eius cremoris optimum concoquens, in ſanguinem id emutat, ſuæ etiam concoctionis duplex recrementum, uti in uini omni que concoctione alia fieri cernimus, obtinens: unum quidem craſſius, quod ueluti ſanguinis fæx & lutū cenſetur, atraq́ bilis uulgò ſerè dicitur: & per portæ uenā in ᵗ lienem ablegatur, qui ad ſiniſtrum uentriculi latus ad inferiora poſterioraq́ repoſitus, linguæ craſſioris imaginem exprimit, accumbentium organorum formæ iecoris ritu congruens, & multis uenis ac item arterijs intertextus, quibus propria iecoris caro lutoſo ſanguini ſimilis, & tenui tunica, quam omentum porrigit, intecta obnaſcitur. Lien itaque craſſius iecoris recrementum ad ſe allicere, atq́ in ipſius nutrimentum conuertere, & ſi quid ſibi adaptare nequit, in uentriculum id eructare cernitur. Tenuius autem iecoris recrementum, quod ueluti flos uini habetur, bilis flaua eſt, quam ᵘ meatus inter portæ & cauæ propagines per iecoris ſubſtantiam digeſti in ſe alliciunt, & ſenſim collecti in unum meatum ceſſant: qui ex iecoris cauo eductus, in bilis ᵘ ueſiculam pertinet, cauæ iecoris ſedi inſtar oblongioris piri media ipſius amplitudine innatam, & corpore conſtantem diſtendi laxariq́ apto. In hac ueſicula bilem aſſeruari diſſectionis profeſſoribus perſuaſum eſt, donec ipſa meatus peculiari beneficio illam in duodenum inteſtinum protrudat: ſimul cum ſiccis uentriculi recrementis egerendam, ac inteſtina ſua mordaci facultate ad propellendum irritatur, atque à pituita illis inſiniſtiſſimum liberaturam. Cæterùm ſanguis dictis nuper excrementis repurgatus, ex anguſtiſſimis uenæ portæ ramis in arctiſſimas cauæ uenæ propagines contendit, tenui aqueoq́ recremento, quod ex inteſtinis in iecur aſſumptum fuit, uehiculi loco utens. Id namque hactenus ſanguinem concomitans, & ſimul cum ſanguine in cauam uenam conſcendens, ſui in his angiportis inſignem uſum præſtat. Quum uerò huc uſque ſanguini promptæ digeſtionis nomine aſtitit, & ſanguis tantam ipſius copiam inſuper non requirit, conſonum fuit, ut id à ſanguine alioqui onus ipſi futurum expurgaretur: cui muneri ˣ renes appoſitiſſimè famulantur, utrinque ſinguli ad uenæ caug latera, & iecori proximè allocati, ac maximam ſeroſi illius humoris portionem ſtrenue ad ſe allicentes, illam que à ſanguine excolentes. Quod quo opportunius moliantur, inſignis uena & item arteria reni tranſuerſim exporriguntur, ac ren in ſinum membraneum & ampliter cauum, in multas que diſſectum propagines, ſeroſum ſanguinem excipit, & ſubſtantiæ renis illi ſinui obnatæ, & duplici tunica obtectæ beneficio urinam expurgat, illam in alium deducens ſinum, quem ʸ urinarius meatus uenæ modo conſtrictus excipit, hanc in ueſicam delaturus. ᶻ Veſica etenim rotundæ lagenæ quodammodo ritu urinam ſenſim admittens, ad poſteriorem pubis oſſis ſedem poſita, peculiari conſtat tunica ſimplici & neruea, triplici que fibrarum genere intertexta, & contrahi ac diſtendi prompta, cui à peritonæo abdominis ue membrana, quæ hactenus dictorum organorum eſt inuolucrum & firmamentum, alia obducitur. In poſteriorem ueſicæ ſedem haud procul ab ipſius ceruice, à ſingulis renibus ſinguli meatus admodum induſtrie inſeruntur, ipſa que tantiſper urinam colligit, donec illa aut copia aut qualitate hominem moleſtans, recluſo orbiculatim ueſicæ ceruicem ambiente muſculo, uniuerſim excernatur. Sanguis hac induſtria repurga repurgatus per cauæ uenæ ramos, ceu riuulos quoſdam uniuerſo corpori digeritur, ut ſingulæ partes ſibi familiare ex ſanguine emulgeant, id que emutantes, & ſibi apponentes, in proprium nutrimentum conuertant: ac demum huius concoctionis quoque recrementa à ſe proprijs functionibus abigant. Porrò ᵇ cauæ ſeries eiuſmo di plurimùm obſeruatur, ubi ea in poſteriori iecoris ſede conſiſt t, ex anteriori ipſius regione ramos promit, in iecoris gibbum numeroſa ſerie digeſtos: ſurſum autem aſcendens, ac ſeptum tranſuerſum cum cordis inuolucro perforans, duas propagines ſepto largitur. Ad dextram cordis aurem caua ampliori orificio in dextrum cordis uentriculum dehiſcit, quàm cauæ orbicularis amplitudo uſpiam uiſitur. A' poſteriori eius implantationis (niſi mauis dicere exortus) ſede, ᶜ uena prodit, cordis baſim coronæ modo cingens, & deorſum per cordis ſuperficiem ramulos digerens. Caua à corde ſurſum conſcendens, ac cordis inuolucrum inibi prætergreſſa, ᵈ uenam paris expertem, & octo frequentius inferiora coſtarum interualla utrinq̃ enutrientem, à dextro latere promit. In iugulo autem tota bipartitò ſcinditur, ab ipſius anteriori ſede pectoris oſſi & interſepientibus thoracem membranis uenas offerens, abdomini ſuperiora perreptantes. A' radice uerò alterius trunci factæ modo in iugulo bipartitionis inſignis promitur uena, ſuper primam coſtam in axillam procurrens, ſed priùs in thoracis cauitate ramum promẽs, in tria ſuperiora ſui lateris coſtarum interualla abſumendum, & alium per tranſuerſos ceruicis uertebrarum proceſſus in caluariam uſq́ue porrigendum, & alium in poſteriora thoracis iuxta ceruicis radicem perreptans. Præſens una thoracæ egreſſa, interdum hic ᵉ humerariam promit uenam, & ramum muſculis anteriori thoracis ſedi inſtratis: alium poſteriori ſedi thoracis, & cauo ſcapulæ: & lateri thoracis alium digerit, in axillam feſtinans. Trunci autem in iugulo factæ bipartitionis reliquum in duos impares ramos diſcinditur, & horum interior gracilior ᶠ internam iugularem conſtituẽs uenam, duabus propaginibus duram cerebri membranam adeuntibus, caluariã ſubit. Exterior uerò ab externo latere propaginem, ex qua humeralis uena conſtituitur, deducens, ſurſum conſcendit, ᵍ ſuperficiariam efformans iugularem uariè ad fauces uſque excurrentem, & in linguam, laryngem, palatum, faciem, tempora & uerticem diſtributam, tribuſq́ uenis caluariam ingredientem. Humeralis uerò priuſq́uam ſub clauicula & ſummo humero in brachium fertur, ramum poſteriori ceruicis ſedi, & alium ſcapulæ gibbo, & alium ſummi humeri ſuperiori ſedi exporrigit, & ſecundum exterius latus muſculi cubitum flectentium anteriori ſub cute prorepens, ſurculoſq́ graciles cuti deprõiens, ſupra cubiti articulum diſſecta, aliquando unum ramum penitius latentem, & mox abſumptum, ad cubiti articulum deſert. alium ſub cute obliquè ad cubiti flexus medium cum ʰ axillaris uenæ ramo coiturum, communemq́ uenam conſtituturum. tertium ſub cute ſecundùm radium in poſteriora cubiti, ac tandem ad brachialis radicem iuxta ulnæ extremum mittit, qui axillaris propagini inibi commixtus, parui & anularis digitorũ exteriora ſcandit. Axillaris in axilla latitans, & cuti anteriorem brachij ſedem uerſus interiora inueſtienti ramum offerens, ſurculum præbet capitibus muſculorum cubitum extendentium, & alium illorum mediæ propemodum longitudini, dein aliam propaginem cum quarto brachium petente neruo ſecundùm brachijpoſteriora ad cubiti uſque exteriora ablegat. Mox in duas ſecta uenas, unam prorſus toto ductu in profundo ſubmerſam, ac arteria ſemper concomitatam, per cubiti articuli flexus medium deducit, quæ ante cubiti longitudinis medium in duas ſecta propagines, unam ſecundũ radium, alteram ſecundùm ulnam uerſus brachiale porrigit. & hic rurſus in ſurculos dirempta, in internam digitorum ſedem ita digeritur, ut ſingulis digitis duæ offerantur ſoboles, & ſurculus quidam inter pollicis primum internodium & poſtbrachialis os indicem ſuſtinens, ad manus externam ſedem pertinget. Altera autem uena ſub cute ſemper exporrecta, iuxta cubiti articulum in duos partitur ramos: quorum alter obliquè uerſus cubiti articuli flexum contendens, cum humeralis ramo cõmiſcetur, ſimul cum illo communem efformans uenam, quæ duabus illis ᵏ medijs uenis conſtituta, ac obliquè deorſum repens, radium'que tandem conſcendens, in externa cubiti ſede inſtar Y in duas diſſcinditur ſoboles, quarum altera medij digiti potiſſimùm externam ſedem petit, altera uerò pollici & indici offertur, ſurculum in internã manus ſedem ſpargens, ramulis montem Veneri ſacrum implicantibus commiſcendum. At alter axillaris iuxta cubiti articulum factæ diuiſionis ramus, uarias propagines in cubiti internam ſedem depromit: quibus crebrò uena accedit, ab altero uenam communem conſtituentium ramo, quem axillaris porrigebat, enata. Hæ propagines admodum

uariè modò coëuntes, modò rurſus inuicem diremptæ, ac internam cubiti ſedis cutem implicantes, tandem in internam manus cutem prorepunt. Cæterùm inſignior eius rami propago ulnæ exporrecta, & in cubiti exteriora etiam ſurculos promens, iuxta brachialis radicem humerariæ ramo commiſcetur, ad paruum digitum & anularem contendenti. Cauæ uenæ pars ſub iecore deorſum repens, à ſiniſtro latere ramum offert pingui renis ſiniſtri tunicæ & locis reni conterminis. Dein ſingulis renibus magna depromitur uena. Ab elatiori ſede uenæ dextrum renem petentis, quæ frequenter altiùs uena ſiniſtro reni propria initiũ ſumit, propago pingui dextri renis tunicam accedit. Ab illius uerò quæ ad ſiniſtrum proficiſcitur renem, humiliori ſede ſiniſtra ſeminalis uena exoritur, dextra interim à cauæ caudice demiſſiùs multo principiũ ducente. Por ub i caua uertebris lumborũ innititur, geniculatim illis propagines dat, in muſculos proximos et ad abdominis uſq́ latera abſumendas. Atq́ harũ illæ ſunt præcipuæ, quæ à caua inibi pronaſcuntur, quà ſupra oſſis ſacriad lumborum uertebras nexum in duos pares diſſcinditur trũcos, quorũ ſiniſter perinde ac dexter ſurculos quoſdam oſſis ſacri foraminibus offert: & in duos partitur ramos, quorum interior propaginem mittit in muſculos exhauriendam, qui oſſis ilium & ſacri poſteriorem ſedem occupant, & aliam deriuat in ueſicam, & penem, & mulieribus in uterũ multiplici ſobole pertinentem. Quod reliquum eius rami eſt, ab externo ramo portiunculam aſſumens, per pubis oſſis foramen in femur ducitur, cuti & muſculis internam femoris ſedem occupantibus ſurculos exhibens, & ante genu articulum ceſſans, ſuoq́ termino cum alterius crus petentis uenæ ramo, ut mox dicam, coiens. Exterior namq́ ramus ſiniſtri cauæ trunci per inguina femur aditurus peritonęo propaginẽ offert, in inferiorem abdominis ſedem ad umbilicum uſq́ abſumendam: in femur uerò procidens, pubis cuti & pudendi muliebris colliculis ſobolem porrigit. Verùm inſignem uenam per femoris & genu & tibiæ interiora, ſub cute ad pedis uſq́ digitorum ſummũ diffundit, in progreſſu ſparſim cuti ramuſculos depromentem. Aliam quoq́ ſub cute ad coxendicis articuli anteriora deriuat. ipſe autem penitiùs inter muſculos ſubmerſus, propaginem communicat muſculis, & cuti in externa femoris ſede locatis, & aliam offert muſculis internã anterioremq́ femoris ſedem ſibi uendicantibus. Atq́ cum hac propagine extremũ uenæ eius coit, quæ per pubis oſſis foramen deſcenderat. Hinc grandis uena ſecundùm femoris poſteriora retorquetur, propagines eius ſedis muſculis communicans, à quibus ramuli in cutem pertinent, tum ſurſum, tum deorſum ad ſuram uſq́ excurrentes. Inter inferiora uerò femoris capita grandis hæc uena in duas ſciſſa truncos, minorem exterioremq́ uerſus fibulam deriuat, à quo præter ſoboles anteriorem genu ſedem petentes ramus diſſcinditur, ſub cute externam tibiæ ſedem uerſus poſteriora obtegente, ad digitorum uſq́ ſuperiora uariè diſſectus. Quod uerò reliquum eſt, altius inter muſculos uenæ fibulæ ſedi attenſos occultatum, ultra tibiæ longitudinis medium properat. Interior autem truncus inſigniter amplus, ramum diſpenſat ſecundùm internam tibiæ ſedem uerſus poſteriora ſub cute ad pedis uſq́ digitos diſperſum. Deinde alium depromit, per ſuram nonnihil reconditum, ad calcem uſq́ue pertingentem. Quod uerò huius trunci præcipuum eſt, muſculos poſteriorem tibiæ ſedem occupantes ſubit, & ab anteriori ipſius regione ramum per ligamentum membraneum demittit, fibulam tibiæ alligans: atq́ is ſub anterioribus tibiam integentibus muſculis abſconditus, ad pedis uſq́ ſuperiora pergit. Ipſa autem ſecundum poſteriora deſcendes, & hinc inde cuti & conterminis muſculis ramulos offerens, tandem inter calcem & tibiam pedis inferiora ſubit, inibi in muſculos & digitos in eum modum diſtributa, ut duæ ſoboles ſingulis offerantur.

DE CORDE, AC ORGANIS
ipſius functioni ſubminiſtrantibus.
Caput IIII.

RGANORVM, quæ inſito nobis calori recreando, & ſpiritibus reſtituendis enutriendiſq́ fabricantur, ᵃ cor iraſcibilis animæ ſedes facile præcipuum cenſetur, inſtar nucis pineæ anterius ac poſterius compreſſæ effigiatum, et ſua baſi ſub pectoris oſſis medio collocatum, ſuoq́ mucrone in ſiniſtrum latus impenſe antrorſum uergens, ſubſtantiaq́ conſtans carnea, ſed quàm muſculorum

E   rum

rum est substantia duriori & triplici fibrarum genere intertexta, ac uenis arterijsq́ue proprijs donata. Dein cor duos[b] sinus seu uentriculos possidet, quorum alter in dextro consistens latere, amplior ac rariori tenuioriq́ue cordis substantia obductus cernitur, & cauæ uenæ orificium ad hunc pertinet, cui tres membranæ introrsum ductæ præficiuntur. Quinetiam uas arteriæ constans corpore, sed uenarum munere fungens, ac proinde[c] arterialis uena appellatum, ab illo uentriculo egreditur, in ipsius orificio tres quoq́ue exigens membranulas extrorsum spectantes. Alter autem uentriculus in sinistro repositus latere, & crassa præcipuaq́ue cordis substantia circundatus, etiam duobus donatur orificijs, quorum humilius uasis cuiusdam est, arteriæ, quod ad aerem attinet, usum præstantis, at uenæ corpore efformati: & proinde[d] uenalis arteria appellati, & in ipsius orificio duas membranas introrsum conniuentes exigentis. Elatius uerò orificium arteriæ magnæ dedicatur principio, cui etiam tres membranas[e] elargitur Natura extrorsum respicientes. Atq́ue his sinus[f] septo interstinguuntur impensè crasso, & comprimi distendiq́ue apto, & in tus foueis plurimis (ut & cordis uentriculi) abundanti corpore extructo. Totum uerò cor membraneo quodam[f] inuolucro tegitur, ipsi nulla ex parte connatum, & corde multo ampliori aqueoq́ue humore intus rorato. Hoc exteriùs inferiori sua sede septo transuerso non mediocri amplitudine connascitur, in lateribus autem duabus[g] membranis cauitatem thoracis intersepientibus, & inuolucrum hoc, ut cor in sua sede seruaretur, suffulcientibus.[h] Pulmo autem reliquam thoracis cauitatem, quam cor & dictæ nuper membranæ & stomachus deorsum descendens non occuparunt, implet, undiq́ue sese accumbentium partium formæ iecoris modo adaptans, ac proinde bubulo, aut alioquin bisulco pedi admodum similis, & primùm dextra parte & sinistra conformatus: quæ rursus singulæ in duas fibras lobos ue diremptæ, multis uasorum implexibus extruuntur.[i] Asperæ namq́ue arteriæ à faucibus (ubi etiam[k] tonsillæ & alia duo glandularum genera consistunt) in thoracem deductæ, & ut uoci famularetur partim cartilagineæ: ut uerò concidere distædiq́ue apta efficeretur, ac respirationi inseruiret, partim membraneæ, rami pulmonè passim implent: & arterialis uena à dextro cordis uentriculo familiarem sanguinè pulmoni præparātè, procedens, cumq́ue sanguinem pulmoni offerens, innumera serie in pulmonem distribuitur, sicuti ac uenalis arteria frequenti serie pulmonem implicans. Atq́ue his uasis fungosa, mollis, spumea & admodum sequax propriaq́ue pulmonis substantia circunfunditur, quam tenuis admodum & pulmonis dilatationem compressionemq́ue haud prohibens tunicula proximè ambit,[l] tunicæ costas succingenti semper attigua. Cæterum quia pulmo thoracis motum è nostro pendentem arbitratu, uacui beneficio sequens dilatatur, uacui etiam ratione extrinsecus nos ambiens aёr per nares secundùm[m] gargareonem: & quum uehementius aёrem ducimus, is per os etiam ueluti in follem attrahiur, ipsius portiuncula per caluariæ foramina cerebrum petens, & reliqua portione per fauces in asperam arteriam subintrans, & pulmonis cauitatem ex ipsius dilatatione factà ad amussim implens. Hunc aёrem pulmonis substantia insita ui alterat, cordisq́ue usibus aptans, optimam sui portionem à uenalis arteriæ ramis undiq́ue asperæ propaginibus attensis assumi permittit, ut eius arteriæ interuentu aёr in sinistrum cordis sinum deferatur, spiritus uitalis idonea futurus materia. Cor enim hunc aёrem attrahens, & in sinistrum ipsius uentriculum magnam sanguinis copiam à dextro alliciens, ex halituoso eius sanguinis uapore, & illo aёre, propria uirtute ipsius substantiæ insita spiritum conficit, quem sanguine impetu ruente, concomitatum fotumq́ue per magnam arteriam uniuerso corpori distribuit, & nasuum cuiusq́ue partis calorem ita temperat, quemadmodum respiratio cor insiti caloris fomitem recreat: itaq́ue idem usus sit respirationis & pulsus, quo arteria magna cordis rhythmo dilatatur & constringitur. Ad spiritum igitur conficiendum cor aere utitur, ipsoq́ue illius feruidus calor temperatur. Verùm quicquid in hac spiritus cõsectione fuliginosum spirituiq́ue efficiendo ineptū est, per uenalem arteriā in pulmonè reduci, atq́ue hinc cū aere qui in pulmone reliquus erat, compresso thorace excerni, dissectionis professoribus est concessum. Adeò sanè ut cor indefessa ipsius dilatatione sanguinem in dextrum ipsius uentriculum à caua attrahat, ut is partim in sinistrū uentriculum ducatur, partim uerò in aptum pulmonis nutrimentū ab ipso amicè præparetur, & contracto corde per arterialem uenam pulmoni offeratur. Cor uerò dilatatum in sinistrum uētriculum ex pulmone aerem assumit, constrictō uerò spiritum uitalem unà cum sanguine impetu ruente in magna arteria propellit. Quò minus autem rapida cordis attractio uenæ cauæ & uenali arteriæ noxam inferret, Natura cordi[n] [...] ures creauit, ueluti promptuaria cordi apponsitas. Quatuor autem cordis uasorum orificijs membranas arbitramur præfectas, ne irritus cordis labor fiat. Membranæ enim cauæ & uenalis arteriæ orificijs præfectæ, impediūt quo minus in cordis contractione sanguis in cauam, & spiritus uitalis in uenalem arteriā refluant. illæ uerò quæ in arterialis uenæ & magnæ arteriæ orificijs habentur, obstant quo minus in cordis dilatatione sanguis pulmoni oblatus, & spiritus uitalis iam emissus, in cor denuò regurgitent. Porrò magna arteria simulatq́ue è corde pronascitur, duas educit[o] propagines cordis basim cingentes, deorsum ramulos per cordis substantiam spargentes. ipse autem arteriæ caudex paulo supra cor in duos sectus truncos, grandiorem sinistrorsum ad spinam detorquet, à quo utrinq́ue ramia ad octo humiliores costas exporriguntur. Quum uerò is sub septo deorsum fertur, huic quoq́ue propagines offert, mox ab una radice omento, uentriculo, iecori, bilis ueficulæ, colo intestino, & demū lieni soboles offerens, uenæ portæ ramis concomitatas. Dein aliam radicem truncus hic in mesenterium dispensat, & utrinq́ue renibus unam deriuat nonnihil inferius, ab ipsius anteriori sede seminales producens arterias, & dein adhuc demissiùs mesenterio alium ramū exhibens, & in progressu uertebris lumborum ipsisq́ue accumbentibus musculis surculos depromens. Ad sacri ossis initium, & si arteria prius caua in sinistro latere subijciatur, eā quo rutior prorepat cōscendit, & eodem cum caua modo hic bipartitò scinditur, parem cum illa ad extremum usq́ue pedis distributionē in profundo faciens. quippe nullus huius arteriæ magnæ trunci ramus cutem subit. Verùm de priuatim huius truci propagini per pubis ossis foramen procedenti accidit, quod illi arteria committitur, quæ ab umbilico secundùm uesicæ latus descendens, fœtui propria censetur. Arteriæ magnæ caudicis truncus superiora petens, statim à sinistro ipsius latere ramum promit, obliquè ad elatissimā thoracis sui lateris costam protensam, à quo primum propago superioribus costis offertur, & dein alia transuersis processibus uertebrarū ceruicis, quæ postmodum in duram cerebri membranam exhauritur. dein alia sinistro pectoris ossis lateri exhibetur, quæ in alto sem per condita, ad umbilicum usq́ue pertingit. Vbi uerò ramus ille thoracis cauitatem superauit, surculum posteriorem ceruicis sedem occupantibus musculis transmittit, & similiter ut axillaris uena ad digitorum usq́ue extrema digeritur, si axillaris ramos cutem subeuntes exceperis, ac in alto latitantes solùm hic intellexeris. Cæterum insignior portio dictīnuper magnæ arteriæ trunci ad iugulū ascendens, in duos impares ramos partitur. sinister, qui & gracilior est, sinistri lateris[p] soporalem constituit arteriam: dexter uerò, ab ipsius dextro latere ad primam costam propaginem derivat, eodem prorsus modo absumendam, quo ramus ille qui obliquè primam sinistri lateris costam adire dicebatur. Verùm dextri rami reliquum, huius lateris soporalem efformat arteriam, quæ similiter atq́ue sinistra secundùm asperæ arteriæ latus fauces petēs, ramum porrigit in faciem penitùs & in temporum cutem ad uerticem usq́ue absorbendum. ipsa autē laryngi & linguæ & triplici hic reposito glandularum generi soboles offerens, caluariam adit, & in duas diuisa propagines, minorem in primum dextrum'ue duræ membranæ sinum exhauriendam mittit. grandior absq́ue uenæ coniugio per proprium foramen in caluariam mergitur, atq́ue hic mox in duræ membranæ latus soboles ab ipsa deriuantur, & alia per peculiare foramen ad narium amplitudinē uersus nasi extremum contendit: ipsa autem propago caluariæ basi instrata, & in nullum tamen plexum reticularè digesta, antrorsum fertur, & ramum cum secundo pari neruorum cerebri ad oculum depromens, sursum ascendit, duram cerebri membranā perforans, & partim in tenuem membranā hic absumpta, partim uerò in dextrum cerebri uentriculum repens, plexus cius in hoc uentriculo repositi, ac extimo fœtus inuolucro comparati portionem efformat, uitalemq́ue spiritum cerebro offert, ut ex illo cerebri beneficio animalis spiritus, uti nunc dicam, præparetur.

DE CEREBRO, ET ORGANIS
cerebri officiorum nomine extructis.
Caput V.

EREBRVM[a] animalis ac principis facultatum sedes in caluaria reponitur, formam cauitatis quam occupat concinnè referens, ac superiori sede secundùm capitis longitudinem, ac anterius posteriusq́ue in dextram partem & sinistram diremptum, in ipsius autem basis medio continuum. Vbi[b] dorsalis medullæ ab ossium medulla plurimùm differentis producit initium. euirūrsus[c] cerebellum

bellum unitur,decuplo ferè minus cerebro, ac pofteriori illius parti omnino fubiectum,neque magis retrorfum cerebrum quàm cerebrum ipfum uergens. Atque hæc uniuerfa cerebri dura amplectitur membrana caluariam proximè fuccinges,& per futuras caluariæ fibras por rigens, quæ in peculiare 'inuolucrum caluariæ degenerant. Hæc membrana tantum à 'tenui cerebri membrana diftat,ut uaforum ipfius motum non præpediat,proceffum mittens inter dextram ce rebri partẽ & finiftrã ,& item aliũ inter fedem cerebelli elatiorem & cerebrum quà cerebello innititur.Huic membranæ quatuor præ cipui infunt finus,uenarum arteriarum'que ufum fimul fubeuntes, & uariam uaforum feriem in tenuem cerebri membranam digeren tes.Cerebri enim fubftantiæ,quæ continua,alba,& nullis uenis in tertexta eft,tenuis quædam membrana uiciniffimè obducitur,quæ paffim in cerebri reuolutiones inteftinorum anfractibus fimilimas fe implicans,uafa cerebri continet. Cerebrum uerò tribus infigni bus et impenfe amplis donatur cauitatibus feu uentriculis,quorum unus fecundùm cerebri longitudinem in dextra cerebri parte confi ftit,quiex pofteriori fede deorfum per cerebri fubftantiam refle xus,ad medium ufque bafis cerebri perfertur.Secundus huic pror fus refpondens,in finiftra cerebri parte collocatur,& ambo quà in ternis lateribus fe inuicem fpectant,fuperiori fede tenui quadam ce rebri fubftantia mutuò diftinguuntur, quam 'feptum nuncupa mus,& quæ fuperius continuatur cerebri portioni,quæ quòd cæ terarum cerebripartium in fuperficie pofitarum fubdurior & ma gis candicans fit, 'callofum corpus appellatur.Humilior autem fep ti fedes unitur continuà'que eft cerebri 'parti quæ inftar fornicis te ftudinis uocatur,utrinque à pofteriori duorum primorum ce rebri uentriculorum fede,ampla bafi pronafcitur,& fenfim antror fum procedens,ueluti in acutum trianguli uerticem coarctatur, hu miliori fua fede quà modò recenfendæ cauitati incumbit, inftar for nicis cauata.Humiliores enim dictorum uentriculorum fedes inui cem fepto non interftinguuntur,uerùm in communem finum con ueniunt,fub corpore inftar fornicis formato repofitum,& uno infi gni meatu rectà deorfum per cerebri fubftantiam pertinentem in fundibulum feu 'peluim à tenui membrana inftar infundibuli ex tructam:quà cerebri pituita per illum meatum defcendens,glandi quadratæ & offi cuneum imitanti fuperftratæ inftillatur,illinc ad palatum & narium amplitudinem per infignia, non autem fpon giæ modo pertufa,foramina defluens. Ifta communis dextri & fini ftri uentriculorum cauitas tertius cerebri eft uentriculus,pofteriori fua fede in meatum definens,qui per cerebri corpora 'teftibus & 'natibus non abfimilia, in quartum uentriculum contendit,qui communis eft cerebello & dorfalis medullæ initio,anteriori ac po fteriori fede'cerebelli 'proceffu ornatus,quem ex anfractuum ima gine uermi in lignis nato comparamus. Verùm in hoc uentriculo, quemadmodum in tribus uentriculis prioribus, nullum peculiare corpus fecantibus occurrit.In dextrum nanque uentriculum (fimi liter ac in finiftrum) per ipfius inferiorem fedem'que conicendit, "plexum efformatu rus,quem extimo fœtus inuolucro comparamus. Is enim ab illo ar teriæ ramo & eius uafis portione conftituitur,quod 'glandulæ uni cis pineæ inftar effigiatæ,& cerebri teftibus incumbenti fuffultum, ab extremo quarti duræ membranæ finus fecundùm cerebelli lon gitudinem exporrecti,per tertium cerebri finuulum ducitur,& ab eo finu tanquam à 'torculari, uenæ arteriæ'que materiam fufci pit.deinceps in duas portiones diffectum,una in dextrum uentricu lum,altera in finiftrum pertinet,ac cum arteriarum ramis cò perue nientibus,dictum nuper plexum in utroque uentriculo conftituit. Ex hoc uitali fpiritu in illo plexucerebri munijs adaptato , ac ex aere quem infpirantes in cerebri uentriculos allicimus , cerebri fubftan tiæ infita uis fpiritũ animalem conficit,quo cerebrum ad principis animæ functiones partim utitur : partim uerò per neruos ab ipfo pronatos, ad organa fpiritu animali indigentia (quæ potiffimùm fenfus ac motus uoluntarij funt inftrumenta) tranfmittit,non me diocrem ipfius portionem à tertio uentriculo fub teftibus cerebri in uentriculum cerebello & dorfali medullæ communem diffundens, quæ poftmodum neruis omnibus à dorfali medulla ortum ducen tibus digeritur.A'cerebri enim bafis medio utrinque unus nafcitur proceffus longus & teres,& fecundùm cerebri bafim antrorfum du ctus,& altero finui octaui capitis offi incumbens,& olfactus orga no proprius:uerùm quoniam ex caluariæ amplitudine non proci dit,nerui nomine à Diffectionum profefforibus non donatus.Pri mum enim feptem quæ cerebro afcribuntur pariũ paulo pofteriùs quàm dicti modò proceffus neruorũ fubftantiæ refpondentes, ini tium à cerebri bafi ducit,uiforios conftituens neruos, qui in oculi

'tunicam retis imagini fimilem degenerant.Oculus namque in cen tro habet 'cryftallinum humorem,cuius anteriori fedi 'tunica te nuiffimæ ceparum pelliculæ correfpondens obnafcitur. Pofterior autem eius humoris regio à 'uitreo humore continetur,cuius pofte riorem ambit fedem inuolucrum cerebri fubftantiæ refpondens, in quod uiforij nerui fubftantia diffoluitur.Tenuis autẽ cerebri mem brana neruum uiforium inueftiens, in 'tunicam dilatefcit uuæ fol liculo perquàm fimilem: quippe ipfa uniuerfum oculum ample ctens, anteriori fedi foramine eò peruia cernitur,quod 'pupillam dicimus. Dura autem cerebri membrana neruo etiam uiforio cir cundata, in 'duram oculi tunicam finitur, toti oculo obductam, & in anteriori oculi cornu modo'pellucidam,quæ 'iride feu maiori oculi circulo circumfcribitur,ad quem 'alba adhærens'ue oculi tunica anteriori oculi fedi obnata definit. Porrò inter corneã hanc & anteriorem cryftallini humoris fedem 'aqueus habetur hu mor,qui à uitreo tunica quadam tenui & ciliorum imaginem expri menti , & orbiculatim cryftallino humori adnata,atque ab uuea principium obtinente diftinguitur.Secundum neruorum par mo uendis oculorum mufculis feruit.Tertium duabus radicibus inui cem diftantibus utrinque enafcens,minorem digerit quadam por tiuncula ad frontis cutem, & quadam ad fuperiorem maxillam fu perius'que labrum, & quadam in narium amplitudinem, & quadã ad mufculos maxillam inferiorem attollentes. Maiorem uerò radi cem tertium hoc par linguæ exporrigit,hęc'que illius beneficio gu ftus inftrumentum efficitur. Verùm ab hac etiam radice ramus ca preoli modo intortus dictis nuper offertur mufculis, & alius fupe rioribus dentibus,alius'que inferiori maxillæ ac dentibus illi infixis, & inferiori tandem labro. Quartum par in palati ceffat tunicam. Quintum duplici quoque radice,ut & tertium par, prodiens,mino rem mufculis maxillam inferiorem attollentibus difpenfat, craffio rem uerò auditus organo offert,quamuis & ab hac binas difpergit foboles per diuerfa foramina in commemoratos modò mufculos e tiam pertinentes. Sextum par præter ramufculos ab ipfo quibuf dam in ceruice mufculis & laryngi in defcenfu oblatos, feptimi pa ris portione adaugetur,atque iuxta pectoris offis fummum fobole quafdam illic produentibus mufculis præbens , ramum cordiarũ radicibus deducit, in organa fanguificationi fubminiftrantia pul chrè digerendum. Ita hactenus uterque fexti pàris neruus pariter di ftributur. Dexter uerò priuatim fui portionem ad arteriam dex tro brachio exporrectam retorquet, à quo neruus confurgit fecun dùm afperæ arteriæ dextrum latus ad laryngem proficilcens, ob id'que neruus 'recurrens nuncupatus. Quod autem reliquum gh hoc dextro neruo defcendit,pulmonis dextræ parti & cordis inuo lucro ramufculos offert, ac ftomacho tandem commiffum, fep tum'que permeans , uentriculi fuperioris orificij finiftram fedem multis propaginibus donat. Siniſter autem ac arteriæ mag næ truncum dorfo explicatum,portiones fuas reuerfium finifter lateris nervum conftituentes retorquet. Ab huius lateris neruo pe culiariter cordi gracilis propago digeritur:quod uraque eius adhuc reftat,dextram fuperioris orificij uentriculi fedem intertexit, fecun dum uentriculi fuperiora ramulum ad iecur utque mittens . Septi mum par præterquàm quod fextum impenfe auget,præcipuè in la ryngis & linguæ mufculos abfumitur. Nerui à dorfali medulla in uertebris conclufa principium ducentes , triginta complectuntur paribus:quorum feptem ceruicis dedicantur uertebris , duodecim thoracis,quinque lumborum, fex facro offi, nullo interim neruo à coccyge offe profiliente. Quæ à ceruicis labuntur uertebris,in mu fculos ipfis adnatos proximos'que digeruntur : & à quarto,quin to & fexto parium furculis utrinque unus efficitur neruus , fepto tranfuerfo proprius:dein à quinto,fexto,feptimo, & dein ab octa uo ac nono,feu primo fecundæ'que thoracis paribus,uarius neruo rum confurgit textus,à quo in brachium fex nerui pullulant, præ ter uarias foboles fcapulæ cauo & gibbo difperfas. Ac primus bra chium petens neruus ab ipfius propaginibus , quas brachium at tollenti offert mufculo,fobolem impenfe gracilem in cutem brachij externæ fedi obductam digerit. Secundus per axillam brachium ingrediens, ramulos'que priori cubitum flectentium mufculo præ bens,infignem ipfius portionem tertio brachium accedenti neruo impartitur.ipfe uerò in cubitum feftinans, & ramulum primo ra dium in fupinum ducenti uniquè foboles,cutem fubit:ac in ua rias diffectus propagines, fuperioris interioris'que cubiti fedis cu tem ad extremam ufque manum implicat . Tertius per axillam quoque defcendens, & anterioris brachij fedis cuti ramulos depre mens,ac fecundi nerui portione adauctus, propaginem'que com municans, pofteriori cubitum flectentium mufculo per anterio

F rem

rem fedem interioris tuberculi offis brachij in cubitum properat, mufculis hinc principiũ ducentibus furculos unã cum quinto neruo fpargens, & fecundùm radium exporrectus, atque in manus uolam ductus, duas foboles pollicis internæ fedi, & totidẽ indici quoque, & unã tantũ externo lateri interioris fedis medij clargitur, non infrequenter etiam medio binos ramulos & unũ anulari offerens. Quartus brachij neruus cæteris multo craffior per axillam ingrediens, & ramos mufculis cubitum extendentibus diffeminans, fecundùm brachij pofteriora ex externum offis brachij tuberculum contendit, in cutem prius duas diffundens foboles. Neruus ad hãc externam articuli cubiti fedem pofitus, ramum in cutem externæ cubiti fedis ad brachiale ufcp fpargens, & mox in duos ueluti truncos diuifus, mufculis'que ab externo brachij offis tubere pronatis foboles dans, unum truncum ulnæ exporrigit, à quo ramuli in mufculos ab externa ipfius fede pronatos digeruntur: ipfe autem truncus iuxta brachialis radicem ceffat. Superior uerò truncus radio exporrigitur, & præter furculos graciles, quos accumbentibus præbet mufculis, brachiale petit, & duos ramulos externæ pollicis fedi, ac duos item indicis, et unum medij digiti interno lateri impartitur. Quintus neruus arteriæ brachij proximus in axilla latitat, & nullas foboles in brachio à fe deducis, in cubitum per poftoriorem interioris tuberculi offis brachij fedem pertingit, ac mufculis hinc pronafcentibus unã cum tertio neruo ramos communicans, fecundùm ulnam ac brachiale excurrit, in medio ductu ramum diffundens, qui duobus furculis in externam parui digiti fedem, & duobus item in anularis, & uno in medij exterioris fedis externum latus abfumitur. Quicquid uerò quinti nerui internam brachialis fedem adit, interiori parui digiti fedi, & anularis & medij ramulos offert. Sextus neruus infigniter gracilis, fecundùm internam brachij fedem fub cute deducitur, & in progreffu ramulos quofdam in cutem diffundens ad cubitum pertingit, in cuius cutem fecundùm ulnam ad brachiale ufque frequentibus furculis diffeminatur. A neruis è thoracis uertebris profilientibus, præter ramos qui retrorfum ad fpinas uertebrarum, & dehinc in mufculos ab illis principium ducentes porriguntur, fingula coftarum interualla fingulos ramos fibi uendicant, ad pectoris & abdominis ufque medium orbiculatim pertinentes, & mufculis thoraci inftratis, abdominis'que mufculis, & demum cuti furculos difpergentes. Ad hæc ab intercoftalibus neruis portiunculæ digeruntur, quæ fexti paris neruorum cerebri propagines coftarum radicibus exporrectis adaugent. Cæterùm diftributio neruorum è lumborum uertebris progredientium thoracis neruis magna ex parte refpondet: illi nanque retrorfum ramos diffeminant, & fecundũ ilia ab abdominis mediũ circulatim afcendunt, ramulos cõterminis mufculis & cuti clargiêtes. Verũ à primo horũ pari ad teftes cum feminalibus arterijs propagines quàm minimæ pertingunt, à quatuor autem humillimis paribus nerui in femur procidentes principium fumunt, quanquam omnium maximus à quatuor primis facri offis neruorum paribus pronafcatur. Primum facri offis neruorum par, fimiliter ac thoracis & lumborum paris, è uertebris labitur. Quincp autem humiliora facri offis paria non à lateribus profiliunt, uerùm una radice antrorfum, altera retrorfum egrediuntur, & pofterioribus radicibus in mufculos facro offi & ilium offibus adnatos & cuti fparguntur. Anterior uerò primipariis ramus unã cum anterioribus triũ fuccedentiũ pariũ radicibus dictum nuper neruũ cõftituit. Humiliorum autem pariũ radices in ueficam, anum & penem, aut mulieribus in uteri ceruice & pudendi colliculos deperduntur. Porro quatuor neruorum in femur progredientium primus fecundùm fextum femur mouente mufculũ deducitur, ac in externã femoris cutẽ ramũ diffeminans, in mufculos abfumitur exterius femoris latus occupantes. Secundus fimul cum grandiori femoris uena & arteria ilium fubit, mox ramum depromens per internam femoris & genu & tibiæ fedem ad fummum ufque pedis digitorum'que extrema fub cute fimul cum uena, quam hâc prorepere dictum eft, defcendens: huc'que ac illuc ramulos digerens. Quod uerò fecundi nerui eft reliquum, in mufculos ceffat anteriorem femoris fedem contegentes. Tertius neruus'pubis offis foramen perreptans, & mufculis id occupantibus ramulos offerens, fobolem deriuat in internam femoris cutem aliquoufcp fparfam: reliquũ autẽ eius nerui in mufculos difcinditur, in interna femoris fede inibi locatos. Quartus omniũ corporis neruorum, qui ex pluribus neruis conftruuntur, facilè craffiffimus, ubi coxendicis os à facro dehifcit, in pofteriora femoris ducitur, ramũ in cutem femoris pofteriorẽ fpargens, qui paulo fub media femoris longitudine ceffat. Humiliori etenim fedi alius offertur ramus, à quarto propagatus neruo, qui etiã mufculis à coxẽdicis offis infima

pofterioticp fede pronatis, foboles exhibet, uti & mufculis ab inferioribus femoris capitibus pronatis. Dein in poplitis regione in duos diffectus truncos, graciliorem exteriorem'que ad fibulam defert, à quo ramus ad externam tibiæ cutem ad paruum ufque digitum repit, & alius in anteriorem tibiæ cutem difpergitur: eius uerò trunci reliquum fibulæ exporrigitur, ubi feptimi & octaui pedem mouentium mufculorum pendet origo. Cæterùm grandior interiorcp truncus ramum in internam tibiæ cutem, & in furæ pariter cutem ad calcem ufque porrigit: ipfe uerò in mufculis furam cõfti tuentibus occultatus, ramum per ligamentum membraneum mittit, quo fibula tibiæ alligatur. hic ramus mufculis anteriorem tibiæ fedem obtinentibus reconditus, tandem in pedis fuperiora contendit, inibi digitis oblatus. Præcipua eius grandioris trunci portio fecundùm tibiæ pofteriora deorfum properans, & nonnullas foboles huc illuc mufculis exhibens, inter calcem interioremcp malleolũ pedis humiliora adit, & exiguas admodum foboles mufculis inibi repofitus offerês, inferiori fingulorũ digitorũ fedi duas propagines cõmunicat. Atcp ad hunc fanè modũ immẽfus rerum Opifex non ad uiuendũ modò, fed etiam ad cõmodè uiuendũ corpus noftrum corruptioni obnoxium condidit, quæ autem ad fpeciei fucceffionẽ fabrefecerit, quicp hæc nutritionis organis procul à fenfibus & rationis fede fubiunxerit, nunc obiter & quantum hæc humanæ fabricæ partium enumeratio admittit, fubijciam.

DE ORGANIS SPECIEI
propagandæ famulantibus.
Caput VI.

INITIO ad fpeciei conferuationem humanæ fabricæ autor duos homines ita exu uxit, ut uir quidem infantis primarium porrigeret principium, mulier uerò id aptè conciperet, infantulumcp ex hoc pronatum principio tantifper nõ fecus quàm aliquod fui corporis membrũ enutriret, quoufque is ualidior redditus, in aërem nos ambiente produci poffet. Atcp his munijs idonea peculiariacp inftrumenta uir ac mulier obtinuit, quibus tanta ad generationem delectationis uis & illecebra inditur, ut animalia hac incitata, fiue iuuenilia, fiue ftulta & rationis expertia fint, haud aliter fpeciei propagandæ incumbant, quàm fi effent fapientiffima. Vir quidem duos adipifcitur' teftes, cute quæ hic° fcortum dicitur, carneã'que menbrana obductos, & alba ac continua peculiari'que prorfus fubftantia extructos: quam' ualida continet menbrana, orbiculatim fubftantiæ huic proximè adnata, ac eorum quæ teftis applantantur infertionem connexumcp excipiens, fingulis'que teftibus peculiare inuolucrũm conftituens. cui° alterum quoque proprium accedit à peritonæo inibi membrana, ubi id uafis feminalibus uiam offert. Illinc namque membrana pronafcitur, ac uafa cum tefte continens, & tefti nufquam, neque etiam feminalibus uafis (nifi quà ex magna peritonei amplitudine excidunt) pertinaciter connexa. Ad humiliorem enim teftis fedem hæc uertebra ipfius carnea parte tantum adnafcitur, quam teftis° mufculum cenfemus. Vafa autem feminalia funt, una utrinque uena, una'que arteria. Vena dextrum teftem petens à cauæ caudicis anteriori fede fub uenarum in renes pertinentium exortu pronafcitur: quæ uiro finiftro tefti offertur, ab humiliori fede uenæ finiftrũ renem adeuntis ob hoc principiũ fumere creditur, ut non purũ fanguinem dextræ uenæ modo tefti perferat, fed ferofum, qui ipfius falfã & acri qualitate in feminis eiectione pruritum concilet. Ambæ arteriæ à grandi arteria paulo inferius quàm dextra feminalis uena initium ducent, & dextra cauæ caudicem confcendens ad dextram fertur uenam, fimul cum hac teftem accedens: & cum uena anteaquam teftem attingat, uarie perplexa, 'corpus'cp efformans multas uarices exprimens, fua'que bafi elatiori teftis fedi infertũ, & ramulos intimo teftis inuoluero offerens, multifariamcp per teftis fubftantiam digeftum, quæ fanguinem hunc benignum, & fpiritum ingenita ipfi facultate in femê haud fecus emutat, quàm iecoris fubftãtia cremorem ab inteftinis ei delatum, in fanguinem alterat. Confectum femen à° ualido uafe inftar uermis teftis pofteriori fedi adnato, uarie'que capreoli ritu implexo excipitur. hoc uas furfum ad magnam peritonæi cauitatem ac uia confcendens, qua feminalis uena & arteria defcendere, nerui cuiufdam modo teres efficitur: & deorfum ad pubis os reflexum, pofteriorem ueficæ fedem accedit, ad quam uas à finiftro tefte femen deferens etiam properat,
quod

quod dextro uafi unitum, fimul cum illo in uefiçæ ceruicis radicem inter[b] glandulofum corpus ceruici hic obnatum inferitur, commu-nifq; femini ac urinæ meatus confurgit, qui deorfum modicé du-ctus, denuò furfum ad pubis offium commiffuram foris reflectitur, corporibus 'penem efformantibus fubnatus . Prodit enim utrinq; à pubis offe unum nerueum teresq; corpus, quod intus impenfe fungofum ac craffo fanguine plenum cernitur. utrumq; fimul con-natum unitumq; penem conftituit, eius fubftantiç beneficio, quum is femen in uterum miffurus eft, erigi augeriq; aptum: et aliàs quum commodum Veneris ufum fuo apice[k] glandis modo tumet, ac' cu-te qua tegi & retegi queat decoratur.   Mulier[m] uterum fibi uen-dicat, femini excipiendo fœtuiq; continendo dedicatur: qui inter ueficam & rectum inteftinum repofitus , uefiçæque inftar fun-do ac ceruice utrifque tendi ac in fe concidere aptis formatus, mem branis laxis & carneis aliquot fibris (quorum auxilio uterus uo-luntarié nonnihil agitur) intertextis , ita fuis lateribus ad perito-næum alligatur , ut mefenterium inteftina continet. Fundi forma non prorfus rotunda exiftit, uerùm anteriùs pofteriusq; depreffa: fuperius obtufa, & duos (utrinque uidelicet unum)[n] retufos often-dens angulos, qui uitulorum fronti cornua producturæ affimilan-tur . Fundo fimplex adeft finus, fundi formæ ad amuffim refpon-dens, & in orificium definens, glandis penis modo in uteri ceruicis amplitudinê prominens, & naturali tantum ui, non autem mulieris arbitratu fefe conftringens & referans. Fundus fimplici propriaq; conftat tunica, nô prægnantibus infigniter craffa, ut utero gerenti-bus in miram amplitudinem extendi ualeat. Huic alia obducitur, à peritonæo initium fumens. Ceruix uteri teres eft, & in utero non ge rentibus aliàs non diftenfa, haud multo minus ipfo fundo ampla, & uefiçæ ceruicis infertionem excipiens,[o] coriaceis'que carnibus &

[c'] colliculis alis'ue ad fuum[o] orificiû ornata. Quinetiam utero utrinq; unus accumbit teftis , ad quem uafa eodem prorfus modo atque in uiris pertingunt. Verùm hic id priuatim accidit, quod media tantù feminalis uenæ arteriæq; pars tefti offeratur, altera uteri fundum im-plicante. Vas femen tenue & impenfe aquofum paucûq; à muliebri tefte deferens, in obtufum fui lateris uteri inferitur angulû. Venæ & item arteriæ frequentiffima ferie, præter antea dictas, uterum impli-cantes, ab illis prodeunt uaforû diftributionibus, quæ poft facri of-fis ad infimam lumborum uertebram nexum perficitur . Atq; hæc uafa enutriêdo fœtui, & ipfius infito calori recreando fubferuiunt. Fœtus autem in utero complexus tribus inuolucris integitur,[r] uno quidem uulgò fecundæ uocato, quod inftar latioris tantum cinguli ipfum ambit , infigniter'que craffum & lienis modo nigricans eft: quod utero connatum, uafa in uterum pertinentia excipit: quo illa in ipfo poftmodû collecta, duabus uenis & totidem arterijs in' um-bilicum inferantur, & una tandem uena iecori, duæ autem arteriæ grandis arteriæ propaginibus per pubis offiû foramina defcenfuris offerantur.[f'] Secundum inuolucrû membrana eft, fœtû uniuerfum amplectens, farciminis'que imagini affimilatum, inter ipfum & tertiû inuolucrum fœtus urinam colligens, quæ peculiari[u] meatu ex uefi-çæ elatiori fede in hanc amplitudinem perfertur, quo minus urina fœtus cuti circumfufa, ipfius acrimonia obfit.[x] Tertium inuolucrû tenuiffima eft membrana, hinc etiam agnina diffectionis proceri-bus appellata, fœtuiq; proximè obducta, ac demum fœtus fudorê inter fe & fœtus cutem fubflauo quafi luto oblitam afferuans. Fœ-tus uerò in lucem editus, lac familiare ipfius alimentum è' mamillis à nemine edoctus fugit: quæ in pectore fedem uendicantes, &[z] pa-pilla ornatæ, glandulofo corpore infita ui fanguinem per uenas ipfi adductum in lac conuertente extruuntur.

(marginal notes left column): b a felucetur eos tuiçar oftit ut. / s lumbit. / k forupt. l uootbu. m priçça. / n lipata. / o vipu.

(marginal notes right column): p utperçà uetra. q serie, uel ticiuet. / r xgiur. / s ζμρφυτ. / t ri ai. readut. / u uirpuçis. / x ζμtet. / y puiçbi. ζ uebi.

ENVMERATIONIS OMNIVM HVMANI CORPORIS FABRI-
cam fubeuntium partium finis: quem quoad fieri potuit, collectiffima integnaq; earundem delineatio in fubfe-
quentibus paginis excipit, ea ferie intuenda, quam mox initio præfcripfimus.

G

ANDREAE VESALII·

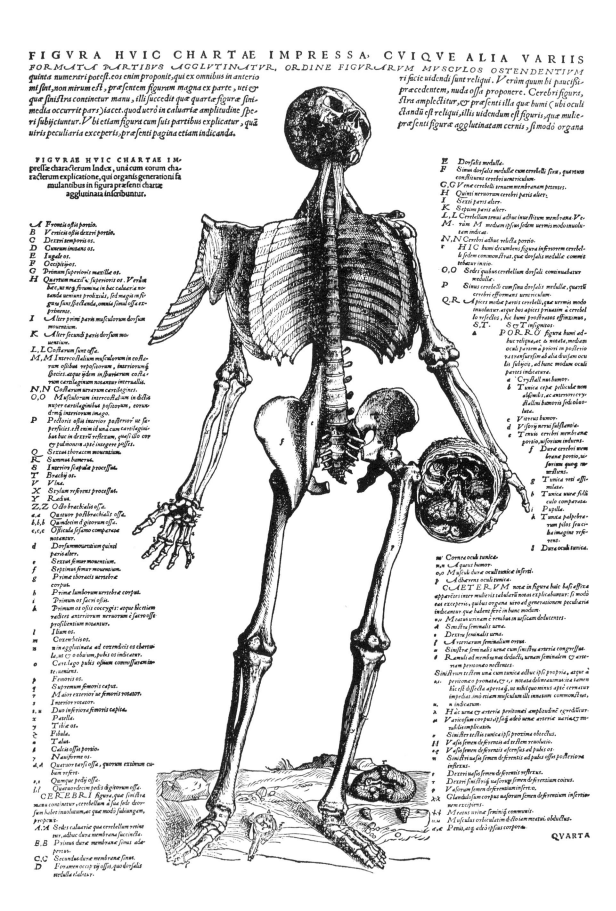

FORMATAE PARTIBVS AGGLVTINATVR, ORDINE FIGVRARVM MVSCVLOS OSTENDENTIVM
quinta numerari poteſt. eos enim proponit, qui ex omnibus in anterio
ri mſunt, non mirum eſt, præſentem figuram magna ex parte, uti ⁊
quæ ſiniſtra continetur manu, illi ſuccedit quæ quartæ figuræ ſini
media occurrit pars) iacet. quod uerò in caluariæ amplitudine ſpe
ri ſubijciuntur. Vbi etiam figura cum ſuis partibus explicatur, quã
uiris peculiaria exceperis, præſenti pagina etiam indicanda.

ri facie uidendi ſunt reliqui. Verùm quum hi pauciſsi
præcedentem, nuda oſſa proponere. Cerebri figura,
ſtra amplectitur, ⁊ præſenti illa quæ humi (ubi oculi
clandũ eſt reliqui, illis uidendum eſt figuris, quæ mulie
præſenti figuræ agglutinatam cernis, ſi modò organa

QVARTA

RI CORPORIS SEDE SVB ILLIS QVI SECVNDAE FIGVRAE SINISTRO LATERE EXPRIMVNTVR
reconditos proponit musculos, dextro quidè latere eos qui primù occur
pes & caput humi collocata pariter iuuant. Tres verò in hac figurae
sequuntur, ac harum prima in integri hominis consistens capite, sinistro
dextra manu hic cōtinetur : illa quae sinistra complectitur, ultimū locum

runt, sinistrò autè qui sub ipsis adhuc cōduntur, quibus ostendendis
conspicuae cerebri partium imagines inuicè sectionis ordine sub
lateri secundae figurae capitis succedit, primam autem excipit quae
sibi uendicante.

SINISTRI LATERIS CHARACTERVM INDEX.

A Capitis processus mamillam referens.
B Tertij paris caput mouentium alter.
C Alter quinti paris caput agentium.
D Alter sexti paris caput agentium.
E Secundi paris dorsum mouentium alter.
F Quarti paris dorsum mouentium alter.
G Octaui paris dorsum mouentium alter.
H Septimi paris dorsum mouentium alter.
I Quinti paris dorsum mouentium alter. Costarum verò osse,
ipsíq; intercostalium musculorum externi, etiam cis a cha
racterum operam conspicientur.
K Scapula prorsus nuda conspicitur.
L Brachij os.
M Vlna.
N Radius. Cæterùm cerebri figura hac manu cūplexa, ce
rebri pars cerebello incūbens ablata est, una cū tata cerebri
portione, ut bases dextri sinistríq; uentriculorum cerebri re
sectæ uideantur. Dein duræ membranæ pars cerebrum ac
O, cerebellum intercedens, ac O & P notata, in sua sede re
P. licta est & eius sinus adapertis sese, uti modò subiungam, os
ferunt.
Q Cerebelli dura membrana non obtecti portio.
R Dexter primus ue duræ membranæ sinus.
S Sinister secundus ue duræ membranæ sinus.
T Dextri sinistríq; duræ membranæ sinuum con
cursus, atq; adeò tertij sinus initium.
V Quintus duræ membranæ sinus.
X Vas à quarto duræ membranæ sinu in tertium
cerebri uentriculum pertinens, hic sursum
reflexirum.
Y Alter cerebri testis.
Z Glandula pineae nucem turbinatamq; figuram
exprimens.
f Cerebri portio.
g Soporalis arteriæ portio, quæ secundum humilio
rem sinistri uentriculi sedem ad plexus secundi
næ forma effici constituentem ascendens.
h Tertij cerebri uentriculi, seu communis dextri
sinistríq; uentriculorum cauitatis portio. Ve
rùm orificium in exteriori sede b conspicuum, e
ius meatus est qui bene pituitam defert. id autem
quod in posteriori h sede apparet, eius meatus
est orificium qui ex tertio uentriculo in
quartum ducitur.
i Os sinistro sacri ossis lateri commissum,
ex carne modò prorsus est.
k Membrana pubis ossis foramen
occupans.
l Femoris os.
m Noni femur mouentium portio.
n Propendet hic cum suis portio
nibus decimus femur mouentium,
a, b, c suos indicans tendines.
o Ab insertione dependet in po
plite latitans musculus.
p Tibiæ os hic nudum modò cernitur.
q Fibula & hic quoq; iam nuda est.
r, r Quintus pedem mouentium.
s Septimus pedem mouentium.
t Octaui pedem mouentium portio.
u, u Carnea moles flexui primorum arti
culorum pedis digitorum præficitur.
x Huius musculi ab insertione pendentis be
neficio quatuor pedis digiti pollici addu
cuntur.
y Secundus pedis digitos mouentium ab in
sertione hic prosternitur.
ʒ Tertius quoq; digitos pedis mouentium hu
mi decumbit.
α Commixtio tendinis tertij musculi cum se
cundo.
Caput hoc, cui sinister pes inniritur, nulla
exprimit occipitium, una cum duabus supre
mis uertebris, ut tertium caput mouentium
β, γ. par β & γ notatum in conspectum ueniret.

DEXTRI LATERIS NOTARVM INDEX.

IN hac cerebrum exprimente figura tantū cerebri ablatum est, quantum elatius cal
loso corpore consistit: ipsum verò callosum corpus in posteriora, bicq; deorsum est remo
latum.
A, A Dextra cerebri pars adhuc reliqua.
B, B Sinistra cerebri pars adhuc reliqua.
G Dexter uentriculus.
D Magna plexus secundinam referentis portio.
E, F, G Corporis instar cameræ testudinis ue efformati supe
rior sedes.
H Euersum hic in posteriora est callosum corpus.
I Linea ab I ad G, rursus ab I ad K pertingens, duorum
K primorum cerebri uentriculorum septum indicat disruptum.
L, M, N, O H is musculosis partibus alter secundi paris mu
sculus constituer caput mouentium.
P Tertij paris dorsum mouentium alter.
Q Quarti paris dorsum mouentium alterius musculi portio.
R Thoracis motorum quartus, a, b, c, d circumscriptus.
S Sexti paris dorsum mouentium alter, a, b circumscriptus.
T Octaui paris dorsum mouentium alterius portio.
V Scapulam mouentium tertius.
X Thoracis motorum tertius, ab insertione hic dependet.
Y Scapulæ gibbum musculis modo liberum conspicitur.
Z Tertium articuli humeri peculiare ligamentum.
f Brachium mouentium tertius.
g Thoracem mouentium secundi portio.
h, h Costarum ossa.
i, i Costarum interualla, atq; adeò exteriores inter
costales musculi.
k Transuersi abdominis musculi portio.
l Brachij os iam hic excarne conspicitur.
m Primus cubitum extendentium.
n Secundi brachium extendentium hic ad
huc asseruata est portio.
o Tertij brachium extendentium exortus.
p, p Radius.
q Vlna.
r Secundus radij in pronum ducentium.
s Ligamentum radium ulnæ qua hæc inui
cem dehiscunt nectens. Quoniam autem
tendines externam manus sedem perrep
tantes resecimus, extremæ manus ossa
ligamentis adhuc cōtecta occur
runt. Cerebri verò figura hæc
manu amplexa, sectionis serie
superiorem subsequitur.
t Corporis cameræ modo ex
tructi inferior cauæq; superfi
cies.
u Sub u uas spectatur à quar
to duræ membranæ sinu in ter
tium cerebri excurrens uentri
culum.
x In dextro uentriculo cerebri plexus secundum referens nota
tur, atq; eadem sinistri lateris est ratio.
y Meatus à tertio cerebri uentriculo pituitam ad glandem huic
excipiendæ idoneam deferentis orificium.
ʒ Magna ex parte ilium ossis dorsum modò nudum conspicitur.
α Tertius femur mouentium.
β Coxendicis articulus.
γ Decimus femur mouentium.
δ, δ Decimum femur mouentium concomitantes carneæ musculo
séq; partes.
ε Ligamentum ab osse sacro in coxendicis os pertinens.
ʃ Magnus exteriori femoris processus.
u, θ, ι, λ Quintus femur mouentium uerum singuli characteres sin
gulas indicant portiones.
n Septimi tibiam mouentium portio.
λ, x Octaui tibiam mouentium portio.
μ Musculus in poplite latitans.
ν Septimus pedem mouentium.
o Octaui pedem mouentium portio.
π Secundus pedis digitos mouentium. Verum huius tendines in
pede hic humi prostrato in suo situ indicantur, eodem chara
ctere insigniti.
ρ Portio tertij pedis digitos mouentium, cuius tendo in pede hu
 σ, mi prostrato etiam ρ & σ indicatur.
τ Tendo est trium primorum pedem mouentium.
τ, υ Tendines secundi pedis digitos mouentium, à sua sede distra
cti, utq; eorum perforatio in conspectum ueniret adhuc asser
uati.
φ, φ Musculi digitos pollici adducentium portiones, qui tendini
bus secundi pedis digitos mouentis exporriguntur.

H TERTIA

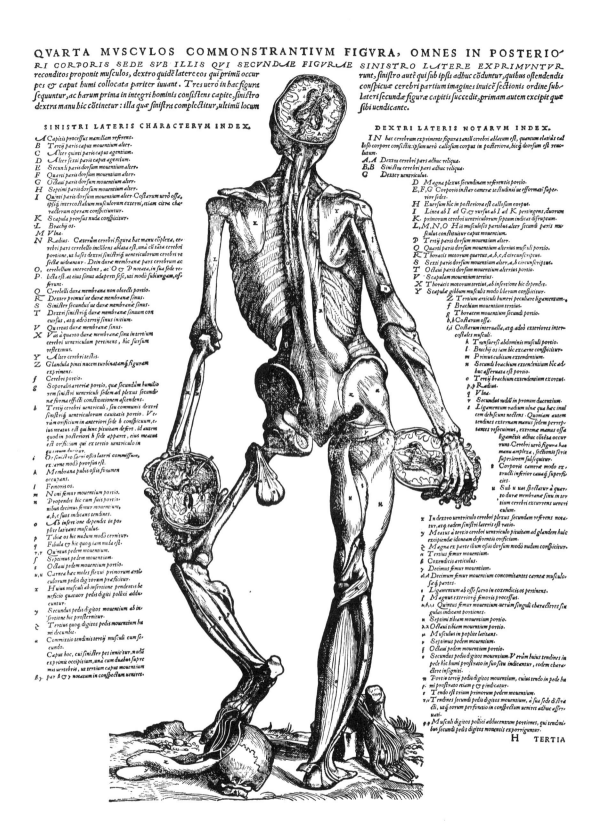

TERTIA MVSCVLOS OSTENDENTIVM FIGVRA, PRIMAM ET SECVNDAM SE
QVITVR, MVSCVLOS ILLIS QVOS IPSAE SINISTRIS LATERIBVS REFEREBANT PROXIME
succumbentes, dextro latere in anteriori facie proponens, dein sinistro
ter linguæ & laryngis musculos, & cartilagines hic humi repositas.
reflexum occurrit.
ater illos referens qui dextri lateris subsunt musculis. præ
Cerebri uerò partium hic nihil propter capitis in posterior

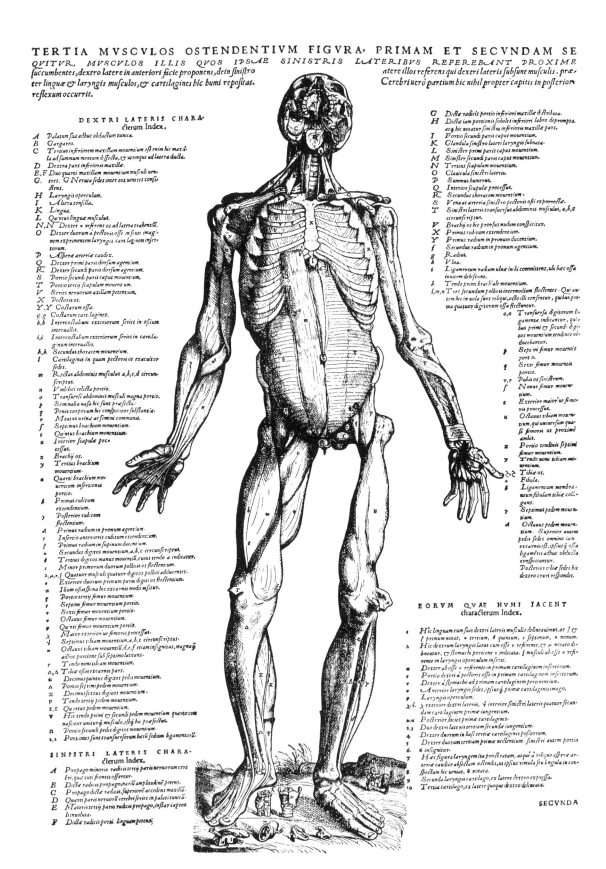

DEXTRI LATERIS CHARActerum Index.

A Palatum sua adhuc obductum tunica.
B Gargareo.
C Tertius inferiorem maxillam mouentium. est enim hic maxilla ad summum mentum dissecta, & utrinque ad latera ducta.
D Dextra pars inferioris maxillae.
E,F Duo quarti maxillam mouentium musculi uersus.
G tres. G'Neruæ sedes inter eos uenies consistens.
H Laryngis operculum.
I Altera tonsilla.
K Lingua.
L Quintus linguae musculus.
N,N Dexter u referens os ad latera trahentium.
O Dexter duorum a pectoris osse in scuti imaginem exprimentium laryngis cartilaginem insertorum.
P Asperæ arteriæ caudex.
Q,R Dexter primi paris dorsum agentium.
R Dexter secundi paris dorsum agentium.
S Portio secundi paris caput mouentium.
T Portio tertij scapulam mouentium.
V Series neruorum axillam petentium.
X Pectoris os.
Y,Y Costarum ossa.
g.g Costarum cartilagines.
b,b Intercostalium exteriorum series in ossium interuallis.
i,i Intercostalium exteriorum series in cartilaginum interuallis.
k,k Secundus thoracem mouentium.
l Cartilaginis in quam pectoris os excauatur sedes.
m Rectus abdominis musculus a,b,c,d circumscriptus.
n Vmbilici relicta portio.
o,p,q Transuersi abdominis musculi magna portio.
q Seminalia uasa hic sunt præsecta.
r Penis corporum hic conspicitur substantia.
s Meatus urinæ ac semini communis.
s Septimus brachium mouentium.
t Quintus brachium mouentium.
u Interior scapulae processus.
x Brachij os.
y Tertius brachium mouentium.
a Quarti brachium mouentium insertionis portio.
ß Primus cubitum extendentium.
y Posterior cubitum flectentium.
δ Primus radium in pronum agentium.
ε Insertio anterioris cubitum extendentium.
ζ Primus radium in supinum ducentium.
η Secundus digitos mouentium, a,b,c circumscriptus.
θ Tertius digitos manus mouentium, cuius tendo a indicatur.
ι Minor primorum duorum pollicis os flectentium.
λ,μ,ξ Quatuor musculi quatuor digitos pollici adducentes.
Exterior duorum primum paris digiti os flectentium.
π Ilium ossis spina hic excarnis modo usitur.
ρ Portio tertij femur mouentium.
σ Septimi femur mouentium portio.
τ Sexti femur mouentium portio.
υ Octauus femur mouentium.
φ Quinti femur mouentium portio.
χ Maior exterior us femoris processus.
ψ Septimus tibiam mouentium, a,b,c circumscriptus.
ω Octauus tibiam mouentium,d,e,f etiam insignius, magnaq́ adhuc portione sub septimo latitans.
Γ Tendo noni tibiam mouentium.
Δ,Δ Tibiæ ossis excarnis pars.
Θ Decimusquintus digitos pedis mouentium.
Λ Portio septimi pedem mouentium.
Ξ Decimussextus digitos mouentium.
Π Tendo tertij pedem mouentium.
Σ,Σ Quartus pedem mouentium.
Υ Hic tendo primi & secundi pedem mouentium quartæ cum nascitur unius musculi, estq́ hic præsectus.
Φ Portio secundi pedis digitos mouentium.
λ,λ Portiones suis transuersorum baru sedium ligamentorum.

SINISTRI LATERIS CHARActerum Index.

A Propago minoris radicis tertij paris neruorum cerebri, quæ cuti frontis offertur.
B Dicta radicis propago, narii amplitudinē petens.
C Propago dictæ radicis, superiorē accedens maxillā.
D Quarti paris neruorū cerebri series in palati tunica.
E Maioris tertij paris radicis propago, instar capreoli inuoluta.
F Dictæ radicis portii linguam petens.

G Dictæ radicis portio inferiori maxillæ distributa.
H Dictæ iam portionis soboles inferiori labro deprompta. atq́ hic notatur sinistra inferioris maxillæ pars.
I,K Portio secundi paris caput mouentium.
K Glandula sinistro lateri laryngis subnata.
L Sinister primi paris caput mouentium.
M Sinister secundi paris caput mouentium.
N Tertius scapulam mouentium.
O Clauicula sinistri lateris.
P Summus humerus.
Q,R Interior scapulæ processus.
R Secundus thoracem mouentium.
S Vena ac arteria sinistro pectoris ossi exporrecta.
T Sinistri lateris transuersus abdominis musculus, a,b,c circumscriptus.
V Brachij os hic prorsus nudum conspicitur.
X Primus cubitum extendentium.
Y Primus radium in pronum ducentium.
f Secundus radium in pronum agentium.
g Radius.
b Vlna.
i Ligamentum radium ulnæ in ibi committens, ubi hæc ossa inuicem dehiscunt.
k Tendo primi brachiale mouentium.
l,m,n Tres secundum pollici internodium flectentes. Qui autem hic in uola sunt reliqui, octo illi censentur, quibus prima quatuor digitorum ossa flectuntur.
o,o Transuersa digitorum ligamenta indicantur, quibus primi & secundi digitos mouentium tendines obducebantur.
p Septimi femur mouentis porto.
q Sexti femur mouentis portio.
r,r Pubis os sinistrum.
ſ Nonus femur mouentium.
t Exterior maior us femoris processus.
u Octauus tibiam mouentium, qui uniuersum qua si femoris os proximè ambit.
x Portio tendinis septimi femur mouentium.
y Tendo noni tibiam mouentium.
z Tibiæ os.
A Fibula.
A Ligamentum membraneum fibulam tibiæ colligans.
γ Septimus pedem mouentium.
A Octauus pedem mouentium. Superior autem pedis sedes omnino iam excarnis est, ipsius q́ ossa ligamentis adhuc obducta conspiciuntur.
Posterior tibiæ sedes hic dextro cruri respondet.

EORVM QVAE HVMI IACENT characterum Index.

ι Hic linguam cum suis dextri lateris musculis delineauimus, ac ? & ſ primum notat, κ tertium, θ quintum, ε septimum, κ nonum.
λ Hic dextrum laryngis latus cum osse u referente, & μ notato delineatur, & stomachi portione u indicata. ſ musculi ab osse u referente in laryngis operculum inserti.
ꝑ Dexter ab osse u referente in primam cartilaginem insertorum.
τ Portio dextri a pectoris osse in primam cartilaginem insertorum.
ꝑ Dexter a stomacho ad primam cartilaginem pertinentium.
φ Anterior laryngis sedes, ipsius q́ primæ cartilaginis imago.
ω Laryngis operculum.
x,ꝗ χ exterior dextri lateris, & interior sinistri lateris quatuor secundæ cartilaginem primæ iungentium.
u,u Posterior facies primæ cartilaginis.
ꝗ,ꝗ Duo secundæ lateris tertiam cartilaginem iungentium.
ꝗ Dexter duorum a basi tertiæ cartilaginis posteorum.
ꝗ Dexter duorum tertiam primæ nectentium sinistri autem portio.
6. 6 insignitur.
7 Hæc figura laryngem ita prostratam, atque a reliquo osse arteriæ caudice abstellam ostendit, ut ipsius rimula seu lingula in conspectum hic ueniat, 8 notata.
8 Secunda laryngis cartilago, ex latere dextro expressa.
9, 10 Tertia cartilago, ex latere quoque dextro delineata.

SECVNDA

SECVNDA COMMONSTRANDIS MVSCVLIS PARATARVM FIGVRA, POSTERIO

REM CORPORIS SEDEM ITA PROPONIT, VT PRIMA ANTERIOREM. IN DEXTRO ENIM LA
tere musculos ostendit cute subtensos, in sinistro autem latere eos qui   illis sectionis serie subijciuntur. Praeterea caput etiam sectionis ordi
ne primae figurae capiti pulchrè succedit.

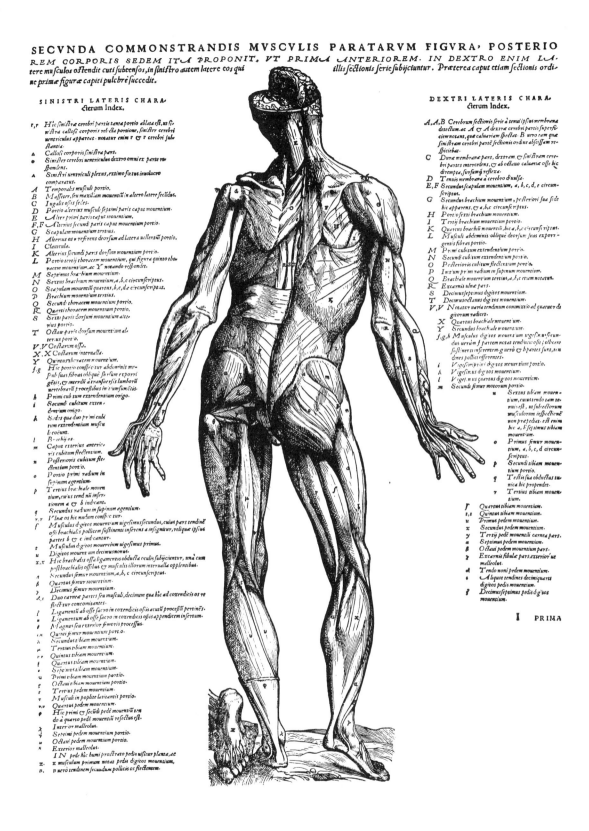

SINISTRI LATERIS CHARA-
cterum Index.

r,r Hic sinistrae cerebri partis tanta portio ablata est, vt si-
     nistra callosi corporis vel ea portione, sinister cerebri
     uentriculus appareat · noteatur enim r & r cerebri sub-
     stantia.
Δ   Callosi corporis sinistra pars.
Θ   Sinister cerebri ventriculus dextro omni ex parte re-
     spondens.
Λ   Sinistri uentriculi plexus, extimo foetus inuolucro
     comparatus.
A   Temporalis musculi portio.
B   Masseter, seu maxillam mouentis in altero latere secudus.
C   Iugalis ossis sedes.
D   Portio alterius musculi septimi paris caput mouentium.
E,F Alter primi paris caput mouentium.
G   Scapulam mouentium tertius.
H   Alterius os a r referens deorsum ad Latera uellentiu portio.
I K Clauicula.
L   Alterius secundi paris dorsum mouentium portio.
M   Portio tertij thoracem mouentium, qui figura quinto tho-
     racem mouentium, ac Y notando respondet.
N   Septimus bra-hium mouentium.
O   Sextus brachium mouentium, a, b, c circumscriptus.
P   Scapulam mouentiu quartus, b, c, d, e circumscriptus.
Q   Brachium mouentium tertius.
R   Secundi thoracem mouentium portio.
S   Quarti thoracem mouentium portio.
     Sexti paris dorsum mouentium alte-
     rius portio.
T   Octaui paris dorsum mouentium al-
     terius portio.
V,V Costarum ossa.
X   Costarum interualla.
Y   ...
Σ   Quintus thoracem mouentium.
f,g Hic portio conspicitur abdominis mu-
     scul: suas fibras obliquè sursum exporri
     gētis, & interdū à transuersis lumborū
     uertebrarū processibus in transumētis.
h   Primi cubitum extendentium origo.
i   Secundi cubitum exten-
     dentium origo.
k   Si d:s qua duo primi cubi
     tum extendentium musculi cocunt.
l   B-echij os.
m   Caput exterius anterio-
     ris cubitum flectentium.
n   Posterioris cubitum fle-
     ctentium portio.
o   Portio primi radium in
     supinum agentium.
p   Tertius brachiale mouen
     tium, cuius tend nū inser
     tionem a & b indicant.
q   Secundus radium in supinum agentium.
r,r Vlna os hic nullum conspic tur.
ſ   Musculus digitos mouen um uigesimus secundus, cuius pars tendinē
     oßi brachialis pollicem sustinenti inserens a insigniatur, reliquae ipsius
     partes b & c indicantur.
t   Musculus digitos mouentium uigesimus primus.
u   Digitos mouent um decimusnonus.
x,x Hic brachialis ossa ligamentis obducta oculis subijciuntur, una cum
     postbrachialis ossiculis & musculis illorum interualla opplentibus.
x   Secundus femur mouentium, a, b, c circumscriptus.
β   Quartus femur mouentium.
δ   Decimus femur mouentium.
ε,ι Duo carnae partes seu musculi, decimum qua hic ad coxendicis os re-
     flectitur concomitantes.
ζ   Ligamentū ab osse sacro in coxendicis os in acutū processū pertinēs.
η   Ligamentum ab osse sacro in coxendicis ossis appendicem inserium.
θ   Magnus seu exterior femoris processus.
ικ   Quintus femur mouentium porto.
λ   Secundus tibiam mouentium.
μ   Tertius tibiam mouentium.
ν   Quintus tibiam mouentium.
ξ   Quartus tibiam mouentium.
ω   Septimus tibiam mouentium.
π   Primi tibiam mouentium portio.
ρ   Octaui tibiam mouentiu portio.
ς   Tertius pedem mouentium.
τ   Musculi in poplite latitantis portio.
υ,υ Quartus pedem mouentium.
φ   Hic primi & secūdi pedē mouentiū ten
     do à quarto pedē mouentiū resectus est.
     Inter or malleolus.
χ   Septimi pedem mouentium portio.
ψ   Octaui pedem mouentium portio.
ω   Exterior malleolus.
     IN pede hic humi prostrato pedis uisitur planta, &
z   μ musculum primum notaet pedis digitos mouentium,
n   η uerò tendinem secundum pollicis os flectemem.

DEXTRI LATERIS CHARA-
cterum Index.

A,A,B Cerebrum sectionis serie à tenui ipsius membrana
       detectum. ac A & A dextrae cerebri partis superfi
       ciem notant, qua caluariam spectat. B uero eam quae
       sinistram cerebri partē sectionis ordine abscissam re-
       spiciebat.
C   Dura membrana pars, dextram & sinistram cere-
       bri partes intercedens, & ab octaua caluariae osse hic
       direpta, sursumque reflexa.
D   Tenuis membrana à cerebro diuulsa.
E,F   Secundus scapulam mouentem, a, b, c, d, e circum-
       scriptus.
G   Secundus brachium mouentium, posteriori sua sede
       hic apparens, & a,b,c circumscriptus.
H   Porto sexti brachium mouentium.
I K L Tertij brachium mouentium portio.
       Quartus brachiū mouentiū, hic a, b, c circumscriptus.
M   Musculi abdominis obliquè deorsum seas expor-
       gentis fibras b.
N   Primi cubitum extendentium portio.
N   Secundi cubitum extendentium portio.
O   Posterioris cubitum flectentium portio.
P   Intum pr:mi radium in supinum mouentium.
Q   Brachiale mouentium tertius, a,b,c etiam notatus.
R   Excarnis ulnae pars.
S   Decimusseptimus digitos mouentium.
T   Decimusoctauus dig:tos mouentium.
V,V   Notatur uaria tendinum commixtio ad quatuor di-
       gitorum radices.
X   Quartus brachiale mouent:um.
Y   Secundus brachiale mouentium.
f,g,h Musculus digitos mouen um uigesimus quartus secun-
       dus uerum f partem notaet tendinis osse oblcem
       sustinent:u inserentem · g uerò & h partes sunt, in
       cines pollicis offerentes.
i   Vigesimūs digitos mouentium porto.
k   Vigesimus dig:tos mouentium.
l   Vigesmus quartus digitos mouentium.
m   Secundus femur motorum portio.
n   Sextus tibiam mouen-
       tium, cum istendo tam te-
       nuis est, ut subiectorum
       musculorum inspectionē
       non praepediat. est enim
       hic a, b sextus tibiam
       mouentium.
o   Primus femur mouen-
       tium, a, b, c, d circum-
       scriptus.
p   Secundi tibiam mouen-
       tium portio.
q   Tertis sua obductus tu-
       nica hic propendet.
r   Tertius tibiam mouen-
       tium.
ſ   Quartus tibiam mouentium.
s,t   Quintus tibiam mouentium.
u   Primus pedem mouentium.
x   Secundus pedem mouentium.
y   Tertij pedē mouentiū carnea pars.
z   Septimus pedem mouentium.
z   Octaui pedem mouentium pars.
A   Excarnis fibulae pars, exterior ue
       malleolus.
Δ   Tendo noni pedem mouentium.
e   Aliquot tendines decimiquarti
       digitos pedis mouentium.
ẜ   Decimusseptimus pedis dig:tos
       mouentium.

I   PRIMA

PRIMA FIGVRA EARVM QVAS OSTENDENDIS MVSCVLIS POTISSIMVM PA

RAVIMVS, IN CVIVS DEXTRO LATERE MVSCVLI MOX SVB CVTE RECONDITI. ANTERIORI IN
facie conspiciuntur: in sinistro autem illis resectis obuij sunt, qui in dextro latere apparentibus proximè succumbunt. Vt uerò præsens
figura redderetur copiosior, eorum quæ caluaria complectitur ima ginem, sectionis ordine proponere incepimus, humi oculi musculos, uti
characterum index docebit, delineantes.

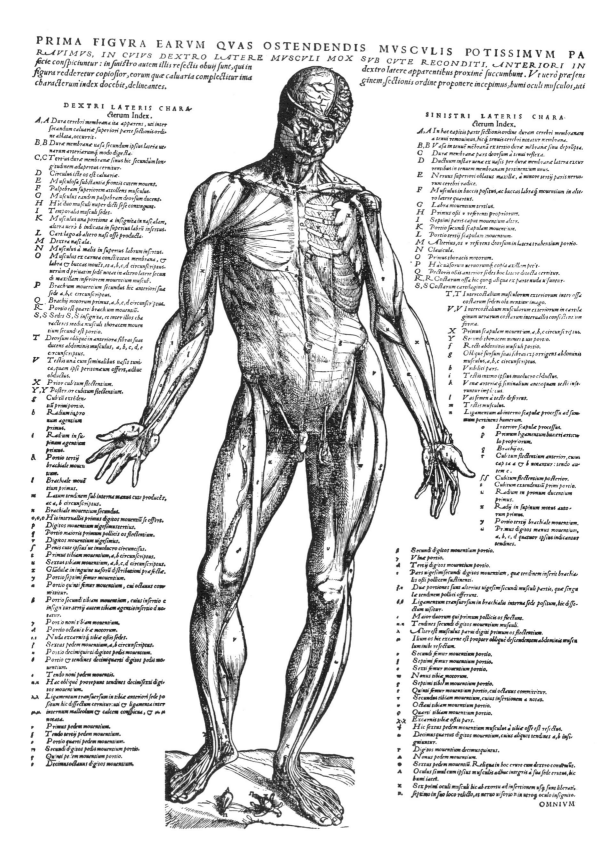

DEXTRI LATERIS CHARActerum Index.

A,A Duræ cerebri membranæ ita apparens, uti inter secundum caluariæ superiori parte sectionis ordine ablata, occurrit.

B,B Duræ membranæ uasa secundum ipsius lateria ue narum arteriarumq; modo digesta.

C,C Tertius duræ membranæ sinus hic secundum longitudinem adspertus cernitur.

D Circulus iste os est caluariæ.

E Musculosa substantia frontis cutem mouens.

F Palpebram superiorem attollens musculus.

G Musculus eandem palpebram deorsum ducens.

H Hic duo musculi nuper dicti sese contingunt.

I Temporalis musculi sedes.

K Musculus una portione a insignita in nasi, alam, altera uero b indicata in superiori labrij insertus.

L Cartilago ab altero nasi osse producta.

M Dextra nasi ala.

N Musculus à malis in superius labrum insertus.

O Musculus ex carnea constitutus membrana, & labra & buccas mouës, et a,b,c,d circunscriptus. uerùm & priuatim hæc notæ in altero latere secun d maxillam inferiorem mouentium muscul.

P Brachium mouentium secundus hic anteriori sua sede a,b,c circunscriptus.

Q Brachij motorum primus, a,b,c,d circunscriptus.

R Portio est quart: brachium mouentij.

S,S Sedes S,S insignia, et inter illos cha racteres media musculi thoracem mouen tium secundæ est portio.

T Deorsum obliquè anteriora fibras suas ducens abdominis musculus, a, b, c, d, e circunscriptus.

V Testis unà cum seminalibus uasis tuni ca, quam ipsi peritonæum offert, adhuc obductus.

X Prior cubitum flectentium.

Y,Y Posterior cubitum flectentium.

g Cubitij extendentium primi portio.

b Radium in pronum agentium primus.

i Radium in supinum agentium primus.

h Portio tertij brachiale mouentium.

l Brachiale mouentium primus.

m Laxum tendinem sub interna manus cute productæ, ac a, b circunscriptus.

n Brachiale mouentium secundus.

o,o,o His interuallis primus digitos mouentiu se offert.

p Digitos mouentium uigesimustertius.

q Portio maioris primum pollicis os flectentium.

r Digitos mouentium uigesimus.

s Penis cutæ ipsius ue inuolucro circuncisus.

t Primus tibiam mouentium, a,b circunscriptus.

u Sextus tibiam mouentium, a,b,c,d circunscriptus.

x Glādulæ in inguine uasorij distributioni præsecldæ.

y Portio sexti femur mouentium.

a Portio quinti femur mouentium, cui octauus committitur.

β Portio secundi tibiam mouentium, cuius insertio e insigni 'ur: tertij autem tibiam agentis insertio d notatur.

γ Portio noni tibiam mouentium.

λ Portio octaui tibie motorum.

ι,ι Nuda excarnis tibiæ ossis sedes.

ſ Sextus pedem mouentium, a,b circunscriptus.

π Portio decimiquinti digitos pedis mouentium.

δ Portio & tendines decimiquarti digitos pedis mo uentium.

ε Tendo noni pedem mouentis.

n,n Hæc obliquè prouepunt tendines decimisexti digi tos mouentiu.

λ,λ Ligamentum transuersum in tibiæ anteriori sede po situm hic dissectum cernitur: ui & ligamenta inter internum malleolum & calcem conspicua, & th et notata.

? Primus pedem mouentium.

ſ Tendo tertij pedem mouentium.

π Portio quarti pedem mouentium.

n Secundi pedem mouentium portio.

? Quinti pedem mouentium portio.

? Decimusoctauus digitos mouentium.

SINISTRI LATERIS CHARActerum Index.

A,A In hac capitis superiori parte sectionis ordine duram cerebri membranam a tenui remouimus, hic & tenuis cerebri notatur membrana.

B,B Vasa in tenui mēbrana ex tertio duræ mēbranæ sinu deprōpta.

C Duræ membranæ pars deorsum à tenui reflexa.

D Ductuum instar uenæ ex uasis per duræ membranæ laterua excur rentibus in tenuem membranam pertinentium unus.

E Neruus superiori oblatus maxillæ, à minore tertij paris neruo rum radice.

F Musculus in buccis positus, ac buccas labraq; mouentium in alte ro latere quartus.

G Labra mouentium tertius.

H Primus ossis u referens propriorum.

I Septimi paris caput mouentium alter.

K Portio secundi scapulam mouentium.

L Portio tertij scapulam mouentium.

M Alterius, os u referens deorsum in latera trahentium portio.

N Clauicula.

O Primus thoracis motorum.

P,Q Hæc uasorum neruorumq; copia axillam pet't.

Q Pectoris ossis anterior sedes hoc latere detecta cernitur.

R,R Costarum ossa hic quoq; aliqua ex parte nuda u suntur.

S,S Costarum cartilagines.

T,T Intercostalium musculorum exteriorum inter ossa costarum sedem obs. mensium imago.

V,V Intercostalium musculorum exteriorum in cartila ginum uterarum costarum interuallis consistentium forma.

X Primus scapulam mouentium, a,b,c circunscrptus.

Y Secundus thoracem mouet s uan portio.

ſ Recti abdominis musculi portio.

g Obliquè sursum suas fibras exp orrigentis abdominis musculus, a,b,c circunscriptus.

b Vmbilici pars.

i Testis intimo ipsius inuolucro obductus.

k Venæ arteriæq; seminalium anteaquam testi inse runtur impl:cus.

l Vas semen à teste deferens.

? Testis musculus.

? Ligamentum ab interno scapulæ processu ad sum uum pertinens humeri.

o Interior scapulæ processus.

p Primum ligamentum humeri articu lo propriorum.

q Brachios.

r Cubitum flectentium anterior, cuius cap ta a & b notantur: tendo au tem c.

ſ,ſ Cubitum flectentium posterior.

? Cubitum extendentij primi portio.

u Radium in pronum ducentium primus.

x Radij in supinum motus auto rum primus.

y Portio tertij brachiale mouentium.

ω Primus digitos manus mouentium, a,b,c,d quatuor ipsius indicantur tendines.

β Secundi digitos mouentium portio.

γ Vinæ portio.

δ Tertij digitos mouentium portio.

ε Pars uigesimisecundi digitos mouentium, quæ tendinem inserit brachia lis ossi pollicem sustinenti.

ſ,n Duæ portiones sunt alterius uigesimisecundi musculi partis, quæ singu le tendinem pollici offerunt.

θ,θ Ligamentum transuersum in brachialis interna sede positum, hic dissectum uisitur.

ι Maior duorum qui primum pollicis os flectunt.

n,n Tendines secundi digitos mouentium musculi.

λ Alter est musculus partu digiti primum os flectentium.

μ Illum ui hic excarne est propter obliquè descendentem abdominis muscu lum inibi resectum.

ν Secundi femur mouentium portio.

ſ Septimi femur mouentium portio.

o Sexti femur mouentium portio.

π Nonus tibiæ motorum.

ſ Septimi tibiæ mouentium portio.

σ Quinti femur mouentium portio, cui octauus committitur.

τ Secundus tibiam mouentium, cuius insertionem a notas.

υ Octaui tibiam mouentium portio.

φ Quarti tibiam mouentium portio.

χ,χ Excarnis tibiæ ossis pars.

ψ Hic sextus pedem mouentium musculus à tibiæ osse est resectus.

ω Decimusquartus digitos mouentium, cuius aliquot tendines a,b insi gniuntur.

? Digitos mouentium decimusquintus.

? Nonus pedem mouentium.

? Sextus pedem mouentij. Reliqua in hoc crure cum dextro construite.

? Oculus simul cum ipsius musculis adhuc integris à sua sede erutus, hic humi iacet.

x Sex primi oculi musculi hic ab exortu ad insertionem usq; sunt liberati.

n Septimo in suo loco relicto, & neruo uisorio n in utroq; oculo insignito.

OMNIVM

OMNIVM HVMANI CORPORIS OSSIVM SIMVL SVA SEDE COMMISSORVM,

OMNIBVSQVE PARTIBVS QVAS STABILIVNT, QVAEQVE ILLIS
adnascuntur, aut ab ipsis prodeunt liberorum integra delineatio.

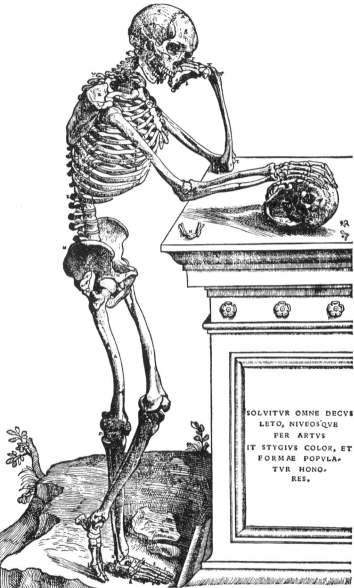

CHARACTERVM HVIVS OMNIVM
ossium imaginem oftendentis figuræ Index.

ET SI in figuris ad præfentem de humani corporis fabrica Epitomen paratis, & ad proportionem uiri mulierisq́; quæ modò subsequuntur, imaginum delineatis, ossa occurrant: quum tamen id passim fiet, neq; eadem figura simul omnia commissis illis partibus nuda detecta spectentur, non abs re fuerit, in hanc paginam aliquin uacantem, ex primo libro huc imaginem reijcere, quæ ex tribus illic omnia ossa simul exprimentibus figuris, hæc magis & anteriori & posteriori sede proponit. Præterquam quod hæc figura etiam caluaræ, cui sinistra manus incumbit, basin una cum osse u imaginem referente & auditus organi osticulis spectandam offerat. V ti characterum Index modò enarrabit.

A Suturarum capitis quæ transuersim feruntur anterior, coronalisq́; dicta.

B Sedes alterius transuersæ suturæ, quæ A imaginem proponit, & à cuius uertice alia ad coronidem, & rarò hinc ad nasi summum fertur, quam à recto ductu sagittalem nuncupamus.

D In temporibus obuia inftar duarum squamarum quæ inuicem incumbunt, commissura. cui ab imagine nomen ind tum est. Reliquis uerò caluariæ superiorisq́; maxillæ suturis, quæ nominibus proprijs destituuntur, nullus adhibitus est character.

* Alterum uerticis os.
β Frontis os nunquam geminum, nisi quam rarissimè à coronali suturæ medio ad summum usque nasi suturam quæ piam pertingit. In huius basi interima. iluariæ amplitudine ossibetur, quod capitis octauum appello.

γ Occipitis os, in grandioribus natu similiter unum.
Δ Dextri temporis os, cuius processus styli aut acus aut galli calcaria imagini comparatus, è insignitur: is uerò qui uberis papillæ conseritur, ς inscriptum ostendit.

ε,* Hæc caluariæ sedes ab imagine & duritia quam cum prærupta rupe similem gerit, nomen obtinet, quod permultis temporum duntaxat ossibus assecribitur.

E Os cuneo à dissectionis proceribus comparatum. Huius processus, quos alijs assimilamus, ξ insigniuntur.

F Hæc caluariæ sedes duorum ossium efformata processibus, perinde ac si peculiare os esset, iugale os nominatur. Superioris autem maxillæ ossa nihil duodenario numero comprehensa. quæ nominibus careat, hæc exigua tabella characteribus non obruenda putatis.

G Inferior maxilla, similiter ac superior sedecim dentibus ut plurimùm decorata.

H Os u imaginem exprimens, unitatisq́; nomine, & si pluribus conflet osticulis, donatum.

ς,* E regione dicti oSH hic iuxta caluariæ basin duo occurrunt auditus organi osticula, quorum alterum q̃ rusignitum molari denti duabus ... radicibus donato, ... imeudi comparatur: alterum ... catum, malleo cuispiam, aut magna ex parte femoris ossi.

I,K,L,M,N His characteribus dorsi ossium seu uertebrarum notatur series, quarum septem ab I ad K pertinentes collum efformant. duodecim à K ad L numeratæ, thoracem. quinque ab L ad regionem M succedentes, lumbos. è regione M, quo ossis sacri in quinta musculorum tabula apparentis, sedes indicatur. ad N sacrum os sex extructum ossibus consistit, cui coccyx os subijcitur.

O Pectoris os pluribus conformatum ossibus.
P Mucronata cartilago, in quam pectoris os gladij modo exacuitur.
1,2,3,4,5,6,7,8,9,10,11,12 Numeri characteribus duodecim costæ indicantur, quarum septem superiores & uertebris & pectoris ossi articulantur, hinc ueræ nuncupatæ: inferiores autem quinque spuriæ dictæ, ad solas uertebras articulationem moliuntur.

Q Clauicula.
R Scapula, cuius superior processus, quem summum humerum dicimus, λ,π is insignitur: interior autem, qui anchoræ imagini conferitur, Σ.
S Brachij os, multis humerus appellatum.
T,V AT ad V sedes metitur, quam cubitum uocemus, duabus constructa ossibus, quorum longius inferiusq́; ac Y insignitum, cubitus priuatim & ulna appellatur. Huius posterior, iuxta illius cum brachij osse articulationem processus v insignitur. Processus autem illius ... cubiti iuxta brachiale conspicuus, ξ inscribitur. Porro superius in cu-
X brio os, radius appellatum, X notatur.
Z Octo brachialis ossa, duplici ordine digesta.
Γ Quatuor postbrachialia ossa.
Δ Quindecim digitorum ossa.
Θ Hoc charactere senistum os eorum quæ sacri ossis lateribus committuntur obuium est. quod in tres sedes, perinde ac si tribus peculiaribus ossibus constitueretur, nomine diuiditur, quæ dextrum os ...
ε,π characteribus notata referet. è nanq̃ ilium os, è coxendicis, ς ... pubis indicat, τ uerò cartilaginem pubis ossi coadunata interuenientem.
Λ Femur, cuius exterior maior & uotator v indicatur, interiori & τ minori interim hic nusquam conspicuo.
Ξ Os scuti patellæ q̃ instar, genu articulo præpositum.
II,Σ AII ad Σ sedes metitur, quam tibiam nuncupamus, quæ q̃ duobus constat ossibus, uno interiori & crassiori, ac tibia appellato, ac * in-
Υ signito: altero graciliori exteriorisq́; quod Υ inscriptum fibulæ nomen obtinuit.
Φ,Χ Inferiorum appendicum ossium quæ tibiam constituunt processus, quos malleolos appellare consueuimus.
Ω Talus tibæ q̃ subditus, calci q̃ inftratus.

a Calcis os, tibiæ rectitudinem longè excedens.
b Os cymbæ imagini à dissectionum peritis comparatum.
c,c Quatuor tarsi ossa, quæ singula characteribus d, e, f, & g in-
d,e dicuntur. ut sic interiora tria quæ nominibus carent, ab extimo,
f,g. quod g insignitur, & tessera è cubo conferitur, distingui queas.

h Quinque ossa tarsum sequentia, pedionq́; constituentia.
i Quatuordecim pedis digitorum ossa.
k Pluribus locis hæc figura k occurrit, osticula indicaturam quæ à sesami semine nomen inuenerunt.

SOLVITVR OMNE DECVS
LETO, NIVEOS'QVE
PER ARTVS
IT STYGIVS COLOR, ET
FORMAE POPVLA-
TVR HONO-
RES.

K EXTERNA.

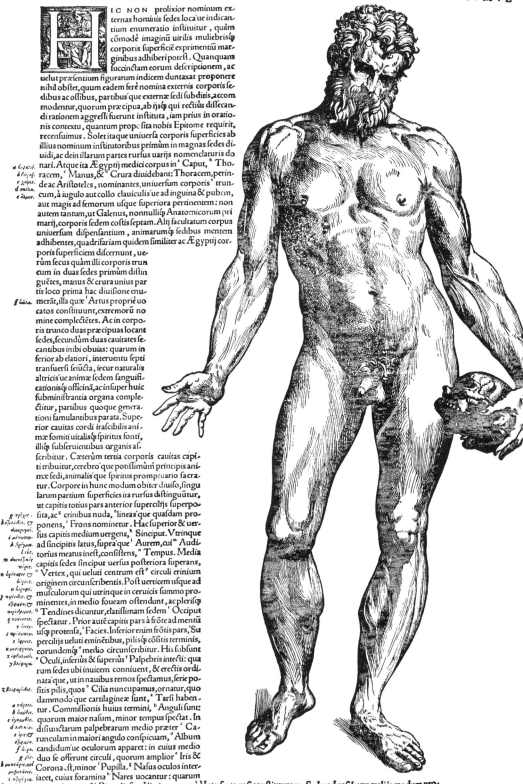

IC NON prolixior nominum externas hominis fedes loca'ue indicantium enumeratio inftituitur, quàm cõmodè imaginũ uirilis muliebris'q corporis fuperficiẽ exprimentiũ marginibus adhiberi poteft. Quanquam fuccinctam eorum defcriptionem, ac uelut præfentium figurarum indicem duntaxat proponere nihil obftet, quum eadem ferè nomina externis corporis fedibus ac offibus, partibus'que externæ fedi fubditis, accõmodentur, quorum præcipua, ab ijs'q qui rectiùs diffecandi rationem aggreffi fuerunt inftituta, iam prius in orationis contextu, quantum propcfita nobis Epitome requirit, recenfuimus. Solet itaque uniuerfa corporis fuperficies ab illius nominum inftitutoribus primùm in magnas fedes diuidi, ac dein illarum partes rurfus uarijs nomenclaturis donari. Atque ita Ægyptij medici corpus in [a] Caput, [b] Thoracem, [c] Manus, & [d] Crura diuidebant: Thoracem, perinde ac Ariftoteles, nominantes, uniuerfum corporis [e] truncum, à iugulo aut collo clauiculis'ue ad inguina & pubem, aut magis ad femorum ufque fuperiora pertinentem: non autem tantum, ut Galenus, nonnulli'q Anatomicorum primarij, corporis fedem coftis feptam. Alij facultatum corpus uniuerfum difpenfantium, animarum'q fedibus mentem adhibentes, quadrifariam quidem fimiliter ac Ægyptij corporis fuperficiem difcernunt, uerùm fecus quàm illi corporis truncum in duas fedes primùm diftinguẽtes, manus & crura unius partis loco prima hac diuifione enumerãt, illa quæ [f] Artus propriè uocatos conftituunt, extremorũ nomine complectẽtes. Ac in corporis trunco duas præcipuas locant fedes, fecundùm duas cauitates fecantibus inibi obuias: quarum inferior ab elatiori, interuentu fepti tranfuerfi feiũcta, iecur naturalis altricis'ue animæ fedem fanguificationis'q officinã, ac infuper huic fubminiftrantia organa complectitur, partibus quoque generationi famulantibus parata. Superior cauitas cordi irafcibilis animæ fomiti uitalis'q fpiritus fonti, illi'q fubferuientibus organis afcribitur. Cæterùm tertia corporis cauitas capiti tribuitur, cerebro'que potiffimũ principis animæ fedi, animalis'que fpiritus promptuario facratur. Corpore in hunc modum obiter diuifo, fingularum partium fuperficies ita rurfus diftinguũtur, ut capitis totius pars anterior fuperciljs fuperpofita, ac [g] crinibus nuda, [h] lineas'que quafdam proponens, [i] Frons nominetur. Hac fuperior & uerfus capitis medium uergens, [k] Sinciput. Vtrinque ad fincipitis latus, fupra'que [l] Aurem, cui [m] Auditorius meatus ineft, confiftens, [n] Tempus. Media capitis fedes finciput uerfus pofteriora fuperans, [o] Vertex, qui ueluti centrum eft [p] circuli crinium originẽ circunfcribentis. Poft uerticem ufque ad mufculorum qui utrinque in ceruicis fummo prominentes, in medio foueam oftendunt, ac pleriif'q [q] Tendines dicuntur, elatiffimam fedem [r] Occiput fpectatur. Prior autẽ capitis pars à frõte ad mentũ ufq protenfa, [f] Facies. Inferior enim frõtis pars, [ſu] Supercilijs ueluti eminẽtibus, pilis'q cõfitis terminis, eorundemꞗ [t] medio circunfcribitur. His fubfunt [v] Oculi, inferiùs & fuperiùs [x] Palpebris intecti: quarum fedes ubi inuicem conniuent, & erectis ordinata'que, ut in nauibus remos fpectamus, ferie fitis pilis, quos [z] Cilia nuncupamus, ornatur, quodammodo'que cartilagineæ funt, [a] Tarfi habentur. Commiffionis huius termini, [b] Anguli funt: quorum maior nafum, minor tempus fpectat. In difiunctarum palpebrarum medio præter [c] Carunculam in maiori angulo confpicuam, [d] Album candidum'ue oculorum apparet: in cuius medio duo fe offerunt circuli, quorum amplior [e] Iris & Corona ſt, minor [f] Pupilla. [g] Nafus oculos interiacet, cuius foramina [h] Nares uocantur: quarum externa latera nafi [i] Pinnulis feu Alis, interna uerò [k] Interfepto nafi conftituuntur. Sedes ad nafi latera mali in modum prominulæ

[Greek annotations in left margins:]
a ἔσχατον.
b θώραξ.
c χεῖρες.
d σκέλη.
e ὅλον.

f ἄρθρα.

g τρίχες.
h ὀυλαὶ, ἢ αἱ αμφυγαί.
i μέτωπον.
k βρέγμα.
l οὖς.
m ἀκουστικός πόρος.
n κρόταφος ὀλιγας.
o κορυφή.
p τρίχωσις. σ στεφάνη. σ ὀφθαλμοὶ πειρόφικε.
q τένοντες. τ ἰνία.
r πρόσωπον.
ſ ὀφρυς.
u μεσόφρυον.
x ὀφθαλμοί.
y βλέφαρα.

z βλεφαρίδες.

a ταρσός.
b κανθὸι.
c ἐγκανθίς.
d λευκόν.
e ἶρις. σεφάνη.
f κόρη.
g ῥίς, μυκτῆρες, uel ῥώθωνες.
h αἱ ῥῖνες.
k διάφραγμα.

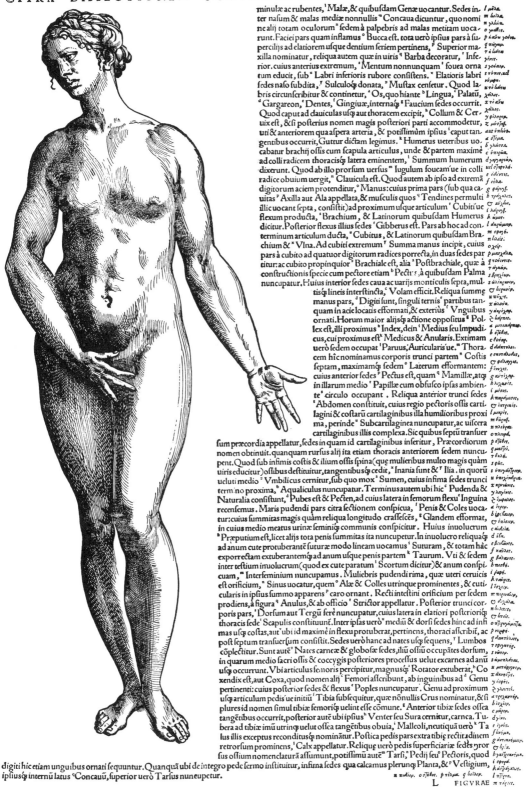

minulæ ac rubentes, 'Malæ, & quibuſdam Genæ uocantur. Sedes inter naſum & malas mediæ nonnullis "Concaua dicuntur, quo nomine alij totam oculorum" ſedem à palpebris ad malas metitam uocarunt. Faciei pars quam inflatum "Bucca eſt. tota uerò ipſius pars à ſuperciliis ad elatiorem uſque dentium ſeriem pertinens, "Superior maxilla nominatur, reliqua autem quæ in uiris "Barba decoratur, 'Inferior. cuius anterius extremum, "Mentum nonnunquam 'ſouea ornatum educit, ſub " Labri inferioris rubore conſiſtens. " Elatioris labri ſedes naſo ſubdita, ' Sulculoſque donata, "Muſtax cenſetur. Quod labris circunſcribitur & continetur, "Os, quo hiante ᵇ Lingua, 'Palatū, ᵈ Gargareon, 'Dentes, 'Gingiuæ, internaque ᵍ Faucium ſedes occurrit. Quod caput ad dauiculas uſque aut thoracem excipit, ʰ Collum & Ceruix eſt, & ſi poſterius nomen magis poſteriori parti accommodetur, uti & anteriorem qua aſpera arteria, & potiſſimùm ipſius 'caput tangentibus occurrit, Guttur dictam legimus. ᵏ Humerus ueteribus uocabatur brachij oſſis cum ſcapula articulus, unde & partem maximè ad colli radicem thoraciſque latera eminentem, ' Summum humerum dixerunt. Quod ab illo prorſum uerſus ᵐ Iugulum ſoueamʼue in colli radice obuium uergit, ⁿ Clauicula eſt. Quod autem ab ipſo ad extrema digitorum aciem protenditur, ᵒ Manus: cuius prima pars (ſub qua cauitas ᵖ Axilla aut Ala appellata, & muſculis quos ᑫ Tendines permulti illi uocant ſepta, conſiſtit) ad proximum uſque articulum ' Cubitiʼue flexum producta, ' Brachium, & Latinorum quibuſdam Humerus dicitur. Poſterior flexus illius ſedes ' Gibberus eſt. Pars ab hoc ad conterminum articulum ducta, "Cubitus, & Latinorum quibuſdam Brachium & ˣ Vlna. Ad cubiti extremum ʸ Summa manus incipit, cuius pars à cubito ad quatuor digitorum radices porrecta, in duas ſedes partitur: ac cubito propinquior ᶻ Brachiale eſt, alia ª Poſtbrachiale, quæ à conſtructionis ſpecie cum pectore etiam ᵇ Pectⁱ s, à quibuſdam Palma nuncupatur. Huius interior ſedes caua ac uarijs monticulis ſepta, multiſque lineis interſtincta, ᶜ Volam efficit. Reliqua ſummæ manus pars, ᵈ Digiti ſunt, ſinguli ternis ᵉ partibus tanquam in acie locatis efformati, & exteriùs 'Vnguibus ornati. Horum maior alijſque actione oppoſitus ᵍ Pollex eſt, illi proximus ʰ Index, dein 'Medius ſeu Impudicus, cui proximus eſt ᵏ Medicus & Anularis. Extimam uerò ſedem occupat 'Paruus, Auricularisʼue. Thoracem hic nominamus corporis trunci partem ᵐ Coſtis ſeptam, maximamque ſedem ⁿ Laterum efformantem: cuius anterior ſedes ᵒ Pectus eſt, quam ᵖ Mamillæ, atque in illarum medio ᑫ Papillæ cum obfuſco ipſas ambiente ' circulo occupant. Reliqua anterior trunci ſedes ' Abdomen conſtituit, cuius regio pectoris oſſis cartilagini & coſtarū cartilaginibus illa humilioribus proxima, perinde "Subcartilaginea nuncupatur, ac uiſcera cartilaginibus illis complexa. Sic quibus ſeptū tranſuerſum præcordia appellatur, ſedes in quam id cartilaginibus inſeritur, Præcordiorum nomen obtinuit. quanquam rurſus alij ita etiam thoracis anteriorem ſedem nuncupent. Quod ſub infimis coſtis & ilium oſſis ſpina (quæ mulieribus multo magis quàm uiris educitur) oſſibus deſtituitur, tangentibuſque cedit, ˣ Inania ſunt & ʸ Ilia. in quorū ueluti medio ᶻ Vmbilicus cernitur, ſub quo mox ª Sumen, cuius infima ſedes trunci termino proxima, ᵇ Aqualiculus nuncupatur. Terminus autem ubi hic ᶜ Pudenda & Naturalia conſiſtunt, ᵈ Pubes eſt & Pecten, ad cuius latera in femorum flexu 'Inguina recenſemus. Maris pudendi pars citra ſectionem conſpicua, 'Penis & Coles uocatur: cuius ſummitas magis quàm reliqua longitudo craſſeſcés, ᵍ Glandem efformat, in cuius medio meatus urinæ ſeminiʼque communis conſpicitur. Huius inuolucrum ʰ Præputium eſt, licet alijs tota penis ſummitas ita nuncupatur. In inuolucro reliquaʼque ad anum cute protuberante ſuturæ modo lineam uocamus ' Suturam, & totam hàc exporrectam exuberantemʼque ad anum uſque penis partem ᵏ Taurum. Vti & ſedem inter teſtium inuolucrum (quod ex cute paratum ' Scortum dicitur) & anum conſpicuam, ᵐ Interſeminium nuncupamus. Muliebris pudendi rima, quæ uteri ceruicis eſt orificium, ⁿ Sinus uocatur, quem ᵒ Alæ & Colles utrinque prominentes, & cuticularis in ipſius ſummo apparens ᵖ caro ornant. Recti inteſtini orificium per ſedem prodiens, à figura ᑫ Anulus, & ab officio ' Strictor appellatur. Poſterior trunci corporis pars, ' Dorſum aut Tergū ferè nuncupatur, cuius latera in elatiori poſteriorique thoracis ſede ' Scapulis conſtituunt. Inter ipſas uerò ᵐ mediū & dorſi ſedes hinc ad infimas uſque coſtas, aut ʼub ubi id maximè in flexu protuberat, pertinens, thoraci aſcribiū, ac poſt ſeptum tranſuerſum conſiſtit. Sedes uerò hanc ad nates uſque ſequens, ' Lumbos complectitur. Sunt autē ˣ Nates carneæ & globoſæ ſedes, iliū oſſiū occupātes dorſum, in quarum medio ſacri oſſis & coccygis poſteriores proceſſus uelut excarnes ad anū uſque occurrunt. Vbi articulus femoris percipitur, magnuſque ʸ Rotator exuberat, ᶻ Coxendix eſt, aut Coxa, quod nomen alij ª Femori aſcribunt, ab inguinibus ad ᵇ Genu pertinenti: cuius poſterior ſedes & flexus ᶜ Poples nuncupatur. Genu ad proximum uſque articulum pediſʼue initiū ᵈ Tibia ſubſequitur, quæ nōnullis Crus nominatur, & ſi plures id nomen ſimul tibiæ femoriʼque uelint eſſe commune. ᵉ Anterior tibiæ ſedes oſſea tangētibus occurrit, poſterior autē ubi ipſius 'Venter ſeu Sura cernitur, carnea. Tubera ad tibiæ imū utrinq́ uelut oſſea tangentibus obuia, ᵍ Malleoli, neutiquā uerò ᵏ Talus illis exceptus recondituſq́ nominatur. Poſtica pedis pars extra tibiæ rectitudinem retrorſum prominens, 'Calx appellatur. Reliquæ uerò pedis ſuperficiariæ ſedes ᵏ ꝓror ſus oſſium nomenclatura aſſumunt, potiſſimū autē ᵐ Tarſi, ᵖ Pedij ſeu ᵖ Pectoris, quod

digiti hic etiam unguibus ornati ſequuntur. Quanquā ubi de integro pede ſermo inſtituitur, infima ſedes qua calcamus plerunq; Planta, & ᵛ Veſtigium, ipſiuſq́ internū latus ᵒ Concauū, ſuperior uerò Tarſus nuncupetur.

QVONIAM figura hic è regione à capite ad proximam paginam glutinata, ac omne uenarum arteriarumque seriem una cum organis tunica costas succingente & peritoneo obductis communi struens, pluribus insignienda uenit characteribus, non abs re fuerit, primùm praecipuas ipsius partes maiusculis Latinis literis indicare, ac dein singula uti casus offeret, alijs characteribus exprimere.

A,A,A Septum transuersum hic conspicitur, quà à costarum et pectoris ossis cartilaginibus recessum est, & dein quà iecoris & lienis posteriora subit.

B Cordis inuolucri est portio mihi relicta, quà id se peò magna amplitudine continuauit.

C Cor iam aptè, ac simplici figura licuit in sua sede expressum, & sua uasa aurei quae ostendens. Huic cordis figurae alias subiecissem, quae ipsius uentriculos, orificia, & ipsis praefectas ostendissent membranas, si modò id una atque altera chartula fieri potuisset.

D,D,D,D Pulmonis quatuor noteantur lobi seu fibrae. Verùm ut cor manifestius una hac figura pingeretur, thoracem inserpsiens membranam omisimus, ac pulmonem uelut in latera reclinatum delineauimus.

E Aspera arteriae caudex simul cum larynge, ipsisque annexis glandulis.

F,F Iecoris gibbi magna superficies tota cum caua se coris sedes sub hac, atque adeò uentriculo delineatur. G & G priuatam insignia.

G,H

I Bilem flauam excipiens uesicula.

I Ventriculus suis sedibus ac neruis, & dein stomacho & superiori membrana omenti, & inferiori etiam humiliori portione ornatus.

K Superior omenti membrana.

L Inferioris membranae omenti portio, quae uelut à colo intestino imam ducit.

M Intestinal ad unum omnia.

N,N Inferioris omenti membranae portio uentriculo subdita, & colum intestinum quà uentriculo ex porrigitur, dorso colligans: atque hic organorum per ipsam & Aristibutorum delineatae series, ut & uasorum distributio per mesenterium in intestina excurrens.

O Lien ex cauda ipsius facie potissimùm expressus.

P Dexter ren.

Q Sinister ren.

R Vterus simul cum suis testibus uasis que delineatus.

S Mulieris uesicula simul cum meatum urinam à renibus deferentium portione, ac dein cum umbilici uasis expressa. Verùm animum modò adhibe notis supra septum positis.

T Venae cauae caudex, quà inter cor & iecur consistit.

V,X Cordis basis, atque adeò uasa id coronariū ritu succingentia.

Y Cordis mucro.

Z Venae sinus ac arteriae per cordis corpus digestae.

a Sedes, qua cauae caudex in cordis dextrum sinum dehiscit.

b Dextra cerebri auricula.

c Sinistrae auriculae apex.

d Venae arterialis caudex. Vualis autem arteriae initium quia in sinistro cordis latere & ut hic cauae orificium in dextro consistit, conspici hac figura nequit. Notant tamen & & uenae arterialis in dextram pulmonis partem progressus, nondum pulmonis substantia undique cir cundatos.

e,e

f Magna arteriae caudex.

g Magna arteriae truncus, partes cordis subdita petens.

h Dicti rami est insigniori portio, atque adeò ipsius in duos impetes ramos distributio, quorum alter sinistra & & soporalis à notata grandior ut & rò dextrum formas soporalem si insignianem, & & teriem potissimùm in dextrū brachium excurrentem, & m notatam.

i Dicti modò trunci insigniori portio, sinistro brachio potissimùm oblata.

k Dicti modò trunci insigniori portio, atque adeò ipsius in duos impetes ramos distributio, quorum alter sinistra & & soporalis à notata grandior ut & rò dextrum formas soporalem si insignianem, & & teriem potissimùm in dextrū brachium excurrentem, & m notatam.

l

m & m notatam.

n,n Hic excurrunt septi transuersi neruiquorum initium in figura cui illa cuius indicem prosequi munt firmauer, P insigniunt.

o Venae distributae intestium, ipsius uerò seriet in praesentis figurae tergo est conspicua.

p Cauae in iugulo dispertitio, ac utrinque ad p Laterae exortus apparent uenarum pectoris ossi propriarum. Harum altera simul cum coniuge sibi arteriae à & ria à insignia, in uno latere & notatam.

q

r Venae superior aliquot costarum interualla adeuntis initium.

---

ff Vena per transuersos ceruicis uertebrarum processus caluariam petens, atque in secundum durae membranae sinum simul cum coniuge arteriam abs sumpta. Notatur enim primus sinus, t,t. secundus x,x.y. dus u, u tertius x,x. quarti initium y.

Axillam petens uena, quae in sinistro latere humeratiam à indicatam promit. In dextro autem latere illius initii hic ab externa pendet iugularis.

b,y Hic abstructae propagines illae sunt, quae ab axillem petente uena in thoracis anteriora, posterioraque latera digeruntur.

A Interior iugularis.

f Interioris iugularis in duas uenas distributio, quarum altera secundum durae membranae peri sinum, altera in latus sinistrum durae membranae sinus.

f Exterioris iugularis.

θ,σ Exterioris iugularis ad fauces distributio. θ ipsius pars post aures ad occipitium excurrens notatur, σ ad tempus σ uerticem, λ ad faciem σ frontem, λ ad secundum durae membranae sinum.

μ uerò duo ipsius rami indicantur, quorum alter per octauum capitis os caluariam petit, alter per secundi paris neruorum cerebri foramen.

Aliquot subsequentes characteres in dextro ponuntur latere, soporalis arteriae seriem indicantes. Atque hic & arteriae portionem indicat caluariam petenem, postquam ramum à sedisse. cum externa iugularis ad faciem, & ad tempus, & post autem digestum.

ξ Soporalis ramus primùm durae membranae petens sinum.

o Praecipua soporalis portio per priuatum foramen caluariam petens.

π Versus narium amplitudinem exporrectus ramus.

ρ Ramus in dextrum latus durae membranae excurrens.

σ,τ Rami soporalis praecipui, qui plexum reticularem efformant perpetram perperam credunt ut.

υ Ramus oculos petens.

φ Ramus tenuem adiens membranam cerebri basi obductam.

χ Plexus, quem extimo foetus inuolucro comparamus.

ψ Sexti paris neruorum cerebri dexter neruus, illic abscissus quà secundum laryngis latus deorsumfertur.

ω,ω Dexter recurrens neruus.

A Sexti paris neruorum cerebri truncus sinister.

B,B Recurrens neruus sinister.

Γ Neruulus cordis basim accedens.

Δ Posterioris cerebri compositi adiens uena.

ε Scapulae posteriora petens uena. Verùm & si posthac arteriam priuatim non insigniamus, prom pti & delineatione liquet, cui iam uenae arteria exporrigatur.

ζ Ad cutem quà summus humerus integitur.

η Humeraria quae cutem subit, cubitumque adit.

θ Humerariae ramus altiora cubiti articuli interdum petens.

ι Ramus ab humeraria ad communem sese constitutionem repens. Verùm modò conuicti duos minores Latinos characteres assumere, ne quam arithmetici characteris geminandi essent, plura ipsis obliterarem.

κ,λ Humerariae ramus insigniori soboles exterioradulnae appendicem & brachiale properans, axil Leuísque ramus p notando, illic ubi n uisitur, aut cla, & paruum digitum, deinde & quae brachiale prae cipuè accedens.

μ Axillaris ramus, cui anteriorem interioremque brachij sedem inuestienti dispensatur.

ν Ad musculos cubitum extendens.

ξ Ramus quartum brachij neruum ad cub ti usque exteriora concomitans.

ο Axillaris in duos truncos partitio.

ππ Truncus in alto latitans, σ latera undique comitatus, σ per cubiti flexum in cubiti tendens.

ρ Exterioris truncus radio exporrectus, σ surculos pollici, indici σ medio exhibens.

σ Arteriae d'idum truncus comitantis soboles, externam manus sedem inter indicem ac pollicem accedens.

τ Reconditi illius trunci ramus ulnae exporrectus, σ paruo digito, aulari σ medio soboles depromens.

υ Sub cutae excurrentis axillaris trunci iuxta cubiti exteriorum diuisio.

φ Axillaris ramus communen constituens.

χ Communis uena.

ψ Communis uena instar Y diuisio, atque dein ipsius per externam manus sedem series.

ο Surculus communis internam manus adiens sedem, σ hic alijs commixtus ramulis.

π Axillaris ramus ulnae exporrectus, uariéque in cutem dissectus, σ suo extremo cum quodam humerariae ramo coiens, ubi n posuimus.

q,q Series uenarum internam cubiti cutem σ uolam implicantium.

γ Vmbilici portio.

σ,σ,σ Vena ab umbilico in iecur exporrecta.

t Meatus urinam foetus inter secundum ipsius inuoluculum σ intimum deferens.

x Vesicae ceruix.

y Vesicae ceruicis musculus.

a,a Superius uentriculi orificium, cui stomachus à tergo agglutinatur figurae continuato. atque hic etiam notatur uena σ arteria coronae modo cū orificium succingens.

γ Stomachi initium cui consulle adhuc affirmatur.

δ Inflexus stomachi ad dextrum latus, ipsíq pecu liares glandes.

Δ Neruorum sexti paris à dextro stomachum series.

θ Vena σ item arteria σ neruus ab elatiori orificio uentriculi al inferius excurrentes. Verùm neruus ille iecur adiens, in iecoris cauo etiam s in dicatur.

μ Inferius uentriculi orificium, σ duodeni intestini initium.

n,n Vena σ arteriae huc ab illis deducta, quae lienis cauo inferuntur.

θ,θ Vasa dextram sedem fundi uentriculi implicantes.

ι,ι Vasa sinistram sedem fundi uentriculi inexentia, σ superiori membranae omenti, uti σ nuper dicta uasa, surculos depromentia.

λ Musculis rectum intestinum simul orbiculatim ambiēs.

λ,λ Musculi rectum intestinum posti egestionem sursum trahentes.

μ Sedes qua colum intestinum recto continuatur, atque ideo interuallum à u ad μ rectum est intestinum.

x,n,n,n Colum intestinum.

ξ,ξ,ξ,ξ Caecum intestinum.

ο,ε,ο Omnes isti anfractus ilei seu uoluuli, intestini ieiuni censentur.

ζ,ω At a usque ad ζ duodeni intestini maxima est portio.

ρ Glandium duodeno intestino, ut σ ieiuni initio adnatum.

σ,σ Meatus bilis uesiculae duodeno insertus.

τ Vena porte caudex.

υ Iecori inserta arteria, σ neruus quoq illi exporrectus arteriae.

φ Bilis uesiculam adiens arteria σ neruus.

χ Vena bilis uesiculam adeuntes.

ψ Vena σ item arteria posteriorem uentriculi sedem iuxta ipsius inferius orificium accedentes.

a Vena uentriculum petens, quà gibbum ipsius dextram dorsi sedem spectat.

b Minor porte maxime distributionis truncus.

c Grandior truncus maxime partitionis una porte.

d Vena σ arteria duodeno intestino potissimùm exporrectae, σ ipsis attento corpore glanduloso sis fultae.

e Vena cum coniuge arteria dextram sedem adiens inferioris membranae omenti.

f,f Radix arteriae in iecur, uentriculum, lienem σ omentum, σ bilis uesiculam digestae.

g Vena σ coniuge arteria superius uentriculi orificium coronae modo tandem cingens, atque in uentriculi figura inter a σ a conspicua.

h,h Vena σ arteria praecipuam inferioris membranae sedem petens, σ colon quà uentriculo exporrigitur implicans.

i,i Glandulosum corpus uasorum distributioni hic praefessum.

k Inferioris omenti membranae sinistram sedem accedens uena.

l,l Vasorum series σ lienem.

m,m,m Vena σ his illis cuae lieni inferuntur sinistram uen tricuti sedem adeuntia. Verùm praecipua in indi cantur, quae sinistram sedem fundi uentriculi im plicant.

o,o,o Series uenarum σ item arteriarum intestinis propriarum.

p,p Minor arteria intestinis propria.

q,q Minor arteria intestini propria.

r,r Glandulae in mesenterio sedem, σ nuper narratos uasorum d'ductus corroborantes.

s Foramen septi transuersi stomachum transmittit

---

teris, atque adeò sinus iecoris stomacho cedens.

t Iecoris ligamentum, quo sinistra ipsius pars septo nectitur.

u Magna arteria septum permeans, atque adeò ipsius ramus in dextram septi partem excurrens.

x Cauae uenae caudex.

y Vena sinistri renis pinguem accedens tunicam.

α Vena σ arteria renis dextrũ oblata.

β Vena dextri renis pinguem tunicam intexens.

γ Sinistram renem accedens uena σ arteria.

δ Sinistra seminalis uena.

ε Dextra seminalis uena.

ζ Arteriarum seminalium ortus.

η Sinistrae seminalis uenae σ arteriae congressus.

θ Ramuli à uena σ arteria in membranam excurrentes, quae peritoneo committuntur.

ι Portio uenae σ arteriae testem adeuntium, superiorem fundi uteri sedem petens.

κ Commixtio uenae σ arteriae seminalium, quae instar pyramidis est, σ uariculas assimilatur.

λ Testis sinister.

μ,μ Vas semen à teste in uterum deferens.

ν Obtusus uteri fundi angulus, in quem uas semen deferens mseruorum molitur.

ξ Hac sede fun uus uterini ceruicem terminatur, haecq, regione ipsius consistit orificium.

ο,π Vteri ceruix.

ρ Hic uesicae ceruix in uteri ceruicem producitur ac desinit.

σ Vasa sunt inferiorum uteri fundi sedem σ cerui cem implicantia.

τ,τ Ceruicis uteri oris colliculi.

υ,υ,υ Meatus urinam è renibus uesicam deducentes.

x Venae σ arteriae lumborum uertebris ipsisq, ad natis musculis, σ abdominis lateribus exporrectae.

a Vena cauae σ arteriae super os sacrum partitio.

c,c Arteriolae sunt sacri ossis foramina petentes.

d Sinistri partitionis dictae trunci diuisio.

e Interioris rami dictae diuisionis propago, nates σ coxendicis ossi adnexa petens.

f Dicti rami propago tandem uesicae σ utero digesta.

g Pars est arteriae foetui peculiaris, quam antea ad uesicae latera u, uti σ hic quoque in maiori figu ra portione, notauimus.

h Portiuncula exterioris rami dictae antea diuisionis ad reliquorum accidens interioris rami.

i Reliquum interioris rami per pubis ossi foramē in musculos interiorem sedem femoris occupantes d'stribuitur.

k Sedes quae soboles dicti reliqui alteri uenae com miscuer. Verùm hic ex tabula promptè animad uertis, quaosità per foramen pubis osis abducendum est, quuos scilicet uenae arteriam attendi cernis.

l Exterioris rami propago, abdominis inferiorem sedem ab umbilico usque perrepens.

m,m Vena per femoris σ tibiae interiora subeat ad digitos usque pedis distributa, σ in progressu na rios sedem surculos.

n Coxendicis anteriora petens, sub cute tamen.

o Musculos σ cutem femoris exteriorem sedem oc cupantes intexentes.

p Femoris anteriorem sedem femoris occupan tius digesta.

q Dictae modò uenae congressus cum ea quae per pu bis osis foramen femur adit.

r Hac praecipua femur petens uena secundum fe moris σ reflectitius.

s,t Propagines musculos posteriorem femoris sedem occupantes, σ dictae suam sedis ad suram usque accedentes.

u Diuisio in poplite, atque sedi σ ramxi in musculos σ femoris capitibus hic pronascentes distributi.

x Maioris dictae diuisionis trunci uena, externam tibiae cutem ad summum usq, sedem implicans.

y Vena σ arteria quanquam non addituum σ qur suis fibulae exporrectae. ac inter musculos latitas.

z Maioris dictae diuisionis trunci ramus, cutem in ternam tibiae sedem integentem, ad digitos usque uarie subiens.

A Ramus dicti trunci suram ad calcem usq, petens.

B Propago grandioris trunci inter musculos tibiae anteriorem sedem occupantes, ad pedis σ superio riora σ digitos digesta.

C Trunci grandioris reliquum inter musculos po steriorem tibiae sedem σ us uendicantes deorsum sum petens, σ inter u biam σ calcem pedis subiens, σ ramulos d'gitorum inferiori sedi cōmunicans.

VENARVM, ARTERIARVM, NERVORVMQVE OMNIVM INTEGRA DELINEATIO,
SIMVL CVM NVTRITIONIS QVAE CIBO POTVQVE FIT ORGANORVM, ET CORDIS PVLMONISQVE
ac demum muliebrium generationi subseruientium instrumento-
rum imaginibus.

FIGVRAE HVIC CHARTAE IMPRESSAE,
cuiᵠ alia agglutinatur, ac quæ peculiariter neruis ostendendis parata est, characterum Index.

CEREBRVM una cum cerebello cerebriᵠ exortibus eum in modum hic
delineauimus, quasi à caluaria erudatum in ipsius basi ita consspiceretur, ut appare-
res si quis erectus caput in posteriora quàm maximè flecteret, sursùm retrorsumᵠ
oculos acturus. Cæterùm quoniam utriusᵠ lateris eadem neruorum in præsenti figura est ratio, suffecerit uni tantùm lateri characteres adhibere.

A,B,C Cerebri ex altero latere basis notatur, ac A indicat partem ipsius ad naraͤ summum prominulum, nonnullis ᵗᵠ mamillarem processum nuncupatam. B uerò cerebri
partem insinuat, amplum caluariæ sinum subeuntem, qui ad latus sinus consistit, quo
glans cerebri pituitam excipiens reponitur. C autem maximè posti-
cam cerebri sedem notat.

D,D Cerebellum.
E Sinister cerebri processus, organo olfactus subministrans.
F Nerui usorij sinistri ortus.
G Neruorum usoriorum coitus.
H Tunica, in quam usorius neruus exoluitur, degenerat ut.
I Secundum neruorum cerebri par.
K,K Minor radix tertij paris.
L Crassior radix tertij paris.
M Quartum par.
N Quinti paris gracilior radix.
O Quinti paris ᵗᵠ signior radix.
P Membrana, in quam quintum par in auditus organo præcipuè
exoluitur.
Q,R Maioris quinti paris radicis propagines, quorum hæc per cœ-
cum elabitur foramen, illa uerò per aliud sibi proprium.
S Sextum neruorum par.
T Septimum neruorum cerebri par, atᵠ horum neruorū progressuᵐ
hic non delineari potuere, quamuis interim magna ex parte in alijs
huius compendij figuris passim, præcipuè uerò in huic superpositis,
ᵗ tertia ad musculos ostendendos præcipuè paratarū occurrant.
V Dorsalis medullæ ex cerebri basis medio initium.
O Dorsal medullæ sedes qua caluariam egrediatur.
1,2,ᵗᵠc Numeri characteres septem ceruicis, duodecim thoracis,
quinᵠ lumborum, ᵗ sex sacri ossis indicāt uertebras, atᵠ adeò triginta neruorum à dorsali medulla prosilientium paria, quorum seriem quàm potui accuratᵉ simᵉ ᵗ simplicissimᵉ in hac tabella meo
marte delineata expressi. Verùm quia hic locus exiguam characterum declarationem admittit, non omnes neruorum soboles literis
insigniturus sum.
P Septi transuersi ᵗ sinister neruus, quem circa aliam characterum
operam ex quarti, quinti ᵗ sexti parium propaginibus efformari
conspicis, promptum enim est delineationem intelligere, si prius de-
scripta neruorum series hîc pictura accommodetur.
Q Neruus à quinto pari cuti summum humerum inuestienti, ᵗ dein
musculo brachium mouenti præcipuè distributus.
R Primus brachij neruus, ipsiusᵠ propagines hic in cutē excurrētes.
S Secundus brachij neruus, ipsiusᵠ in anteriorem cubitum flectentium musculum soboles.
T Tertius brachij neruus, ipsiusᵠ ᵗ propago cuti anteriorem brachij
sedem induenti oblata.
V Tertij nerui propago ad musculum posteriorē cubiti flectentiū.
X Secundi nerui portio tertiæ accedens.
Y Ramus secundi caput adiens longioris radium in pronum mouentis musculi.
Z Secundi distributio in duos impares ramos.
a Minor ramus secundum radium cuti ad pollicē usᵠ exporrectus.
b Crassior ramus mox in duas propagines diuisus, quarum series in
conspicuo est.
c Tertij nerui soboles in musculos internam cubiti sedem occupantes digestæ.
d Tertij ramus radio exporrectus, ac dein pollici ᵗ indici ᵗ medio
surculos offerens.
e,e Quartus brachij neruus atᵠ inferius ᵗ ramos notas musculis cubitum extendentibus depromptos.
f Quarti ramus internam brachij cutem adiens.
g Quarti ramus externam ᵗ posteriorem brachij accedens cutem.
b,b Quarti ramus cuti externæ cubiti digestus.
i Quarti præcipua distributio ad ingrestuͤ cubiti.
k,k Quarti ramus radio exporrectus, ᵗ externæ sedi pollicis, indicis,
ᵗ medij soboles implicans.
l,l Quarti ramus ulnæ exporrectus, ᵗ musculis ab externa ipsius sede initium ducentibus ramulos exhibens, ac ante brachiale cessans.
m Quintus brachij neruus.
n Quinti nerui series in musculos ab interno brachij ossis tubere pronatos.
o,o Quinti nerui ulnæ exporrectus, ᵗ internæ sedi parui digiti
ᵗ anularis, aliquandò ᵗ medij ramusculos dispergens.
p Dicti modò rami soboles in externam manus sedem reflexæ, ac externæ parui digiti sedi ᵗ anularis ᵗ medij surculos dispensans.
M Sextus brachij neruus, ipsiusᵠ sub cute tantù ducta series. Quis
uerò brachij neruorum sit exortus, principiorumᵠ plexus, alᵗᵠ
notis promptè dignoscitur.
r,r,r Nerui sunt intercostales, illic præcisi, quà cum costis antrorsum reflectuntur.
s,s Rami in posteriora deducti, hic etiam undiᵠ obuij.
t,t Huiusmodi serie nerui musculos adeunt thoracis ossibus instratos.
u,u Propagines indicantur, sexti paris neruorum certium com-
flarum radicibus exporrectum augentes.
x,x Propagines neruorum ex lumborum uertebris exilientium, quæ abdominis ᵗ huius sedis musculis ᵗ cuti dispensant.
y Neruulus testem frequenter petens, hicᵠ resectus.
z Propagines sextum femur mouentium adeuntes musculum.
σ Primus femur petens neruus.

α Primi nerui propago cuti oblata.
β Primi nerui propago altius inter musculos absumpta.
γ Secundus femur petens neruus.
δ,δ Secundi nerui soboles per internam femoris ex-
tibiæ sedem ad pedis usᵠ superiora sub cute excurrens.
ε Secundi nerui propago musculis anteriorem femoris sedem occupantibus depromsta.
ζ Tertius femoris neruus.
η Tertij propago, internam femoris cutem implicans.
θ Tertij propago musculos adiens.
ι,ι Quartus femoris neruus, cuius exortus æquè
atᵠ trium superiorum est conspicuus.
κ,κ Series anteriorum propaginum inferiorum parium è sacro osse productarum.
λ Dorsalis medullæ extremum.
μ Quarti femur petentis nerui propagines ad capita musculorum à coxendicis ossis appendice
productarum sparsæ.
ν Quarti nerui soboles in posteriorē femoris cutē
ad medium usᵠ longitudinis femoris excurrēs.
ξ Propago præcipua in quartum tibiam mouentium musculum, ac dein in reliquam posteriorem
femoris cutem iuxta genu digesta.
ο Soboles in musculos ab inferioribus femoris capitibus pronatos.
π,ρ Quarti nerui in duos truncos distributio, ac ρ
quidem minorem, ρ uerò insigniorem notat.
ς Minoris trunci propago, externæ tibiæ cutē
ad parui digiti usᵠ extremum diffusa.
τ Propago fibulæ inter musculos exporrecta.
υ Ramulus anteriorum tibæ cutem implicans.
φ,ψ Grandioris trunci ramus internæ cuti tibiæ ad
pollicem usᵠ digestus.
χ Grandioris trunci neruus, posteriori tibiæ seu
suræ musculos usᵠ digestus.
ψ Grandioris trunci ramus, ligamenti penetrās
fibulam tibiæ quà hæc ossa inuicem dehiscunt cō-
mittens, ac dein ad digitorum usᵠ finem
excurrens.
ω Præcipua grandioris trunci portio, inter tibiæ
os ᵗ calcem pedis inferiora petens, singulisᵠ
digitis surculos offerens.

FINIS.

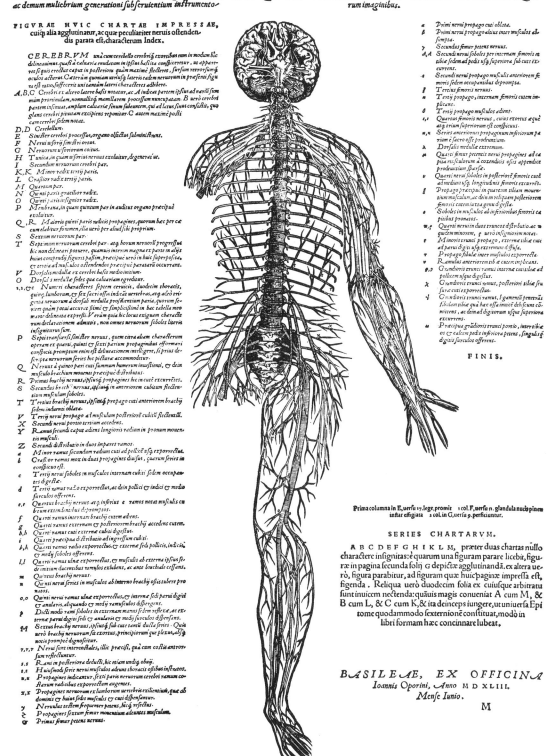

Prima columna in E, uersu 17. lege, promit 1 col. F, uersu n. glandula nuclᵉ pineæ
instar effigiata 2 col. in G, uersu 9. perficiuntur.

SERIES CHARTARVM.

A B C D E F G H I K L M, præter duas chartas nullo
charactere insignitas: è quarum una figuram parare licebit, figuræ in pagina secunda folij a depictæ agglutinandā. ex altera uerò, figura parabitur, ad figuram quæ huic paginæ impressa est,
figenda . Reliqua uerò duodecim folia ex cuiusque arbitratu
sunt inuicem nectenda: quauis magis conueniat A cum M, &
B cum L, & C cum K, & ita deinceps iungere, ut uniuersa Epitome quodammodo sexternione constituat, modò in
libri formam hæc concinnare lubeat.

BASILEAE, EX OFFICINA
Ioannis Oporini, Anno M D XLIII.
Mense Iunio.

M

AD figuram neruorum feriem in ultima totius Epitomes charta proponentem, figura multis extructa partibus glutinatur, quæ præter uenarum arteriarumq́ feriem, & nutritionis quæ cibo potuq́ fit organorum, & cor partiumq́ ipfi fubminiftrantuli imagines, generationis in mulieribus inftrumenta proponit. Verùm quia uiri alicubi fpectanda erant, præfentè fecī chartam parauimus, quæ ab illa quæ figurarum ultimæ chartæ committendam continet, nulla ex parte differt, præterquam in generationis organis. Licet itaque pleræque in hac charta impreſſæ figuræ illis refpondent quæ in nuperdicta proponuntur charta, neutiquam obfuerit, ex bis omnibus unam parate tabulam, quam quinte mufculorum figuræ affiges. Totiitaque chartæ membranam fubglutinabis, omnesq́ figuras à fuperflua papyro circumfcinde, nifi quòd in prima quæ cæterarum eft præcipua, ac uelut aliarum bafis, fupra caput portiunculam feruari conuenit, à qua tota agglutinata pendeat. Dein unæ ac arteriæ ad u.ficam & pudenda excurrentes, ab abundanti papyro non funt refecandæ, ut QVARTA figura quæ inteftina proponit ipfi conglutinari illic queat, ubi n infcripti habet, nifi fortè illam SECVNDAE quæ uentriculum, ftomachum & mefenteriū fuperiorem membranam proponit: tergo iungere uelles, ut utriufq́ figuræ ſ refponderet. TERTIA humilioris membranæ, quæ fub colo inteftini parte uentriculo extorrecta confiftit, omenti poftériorem faciem oftendens, ad fecundam ita nectetur, ut K opponatur L, ac facculi imago infurgat. Nunc fecundam ad primam nexu us, in fepto tranſuerfo ubi ſ occurrit, fecurè ftomacho cedis foramen molieris, per quod ftomachum ita traijcies, ut afferere arteria fuccumbit, nexusq́ ad foramini stergum fiat. QVINTA humilioris n̄ embranæ omenti portionem uentriculo pofteriori fedi fubditam cum liene & portæ uenæ illi quæ afterfis arterijs diftributione commonſtrans, cauæ iecoris fedi in prima figura gluti nal:tur, ubi in ambabus figuris u, ʃ, x, ſ occurrunt. SEXTA, quæ meatuum urinam è renibus in uesicam deferentiū portionem exprimit, & feminalium uaforum cum teftibus ipforumq́ inuolucris commonſtrat, illic primæ eft iungenda, ubi uenæ & arteriæ feminales urinæ vijs meat:bus incumbere uidentur, aut ubi in fini ſtro latere in utraq́ue figura n ſpectatur. SEPTIMA figura infériorem penis corporū fedem & meatum urinæ feminiq́, cōmunem cum ipfum ambiente mufculo oftendens, ex proportione fub octaua eſt conglut:nada, ut ω, ε, ʃ, muicè refpondeant. OCTAVA autem uesicam, umbilicum, cum ipfius uasis & glandulofo corpore uesicæ ceruici adnato & pene commonſtrans, ad feptam illic eft committenda ubi ϕ ſpectatur, ut fcilicet ϕ feptimæ figuræ ad ϕ fextæ committatur, ac dein penem infter S litere complices. NONA portionem gibbæ fedis iecoris proponens, non inutiliter ueluti ex puncto illic glutinabitur, ubi A inter ʃ & F maioris feu primæ figuræ confiftit. DECIMA, quæ uenæ pari carentis diſtributionem proponit, magnæ figuræ tergo illic uenit committenda, ubi cauæ caudex eam uenam promit, ac o in utrifq́ figuris fcripum ſpectatur. VNDECIMA, ubi à redundanti papyro circumcifa erit, duas partes conftituet, quarum fuperior uenam arteriamq́ dextri lateris exprimit, fub pectoris offe fuperiorem abdominis fedem petentes. Huius itaque partis q ad magnæ figuræ q figetur, & * ad ramum qui ad dextrum latus ϛ & m in magna figura occurrit. Ceterùm humiliori pars uenam arteriamq́ exprimens, quæ infériorem abdominis fedem explicant, illic eft iungenda, ubi earum radices iuxta l in dextro latere ſpectetur.

Figuræ quæ DVODECIMA infcribitur, & ʃ huic chartæ impreſſæ fit, non ad primam figuram eſt committenda, uerùm ad illam quæ quintæ mufculorum tabula uocatur. Hic enim pubis oſtium anterior fedes cum cartilagine oſtium illorum coalitum interuenien te, delineatur. Qui autem glutinanda fit, promptiſsimè obſeruabis, quum illam à fupẽrflua char ta recifam, altera ex parte quintæ mufculorum tabulæ committere ſtudebis, ubi pubis oſſa ab illa utides adempta.

SECVNDA.

PRI-MA.

DVODE-CIMA.

SEPTIMA.

OCTAVA.

TERTIA:

QVARTA.

VNDECIMA.

NONA.

QVINTA.

DECIMA.

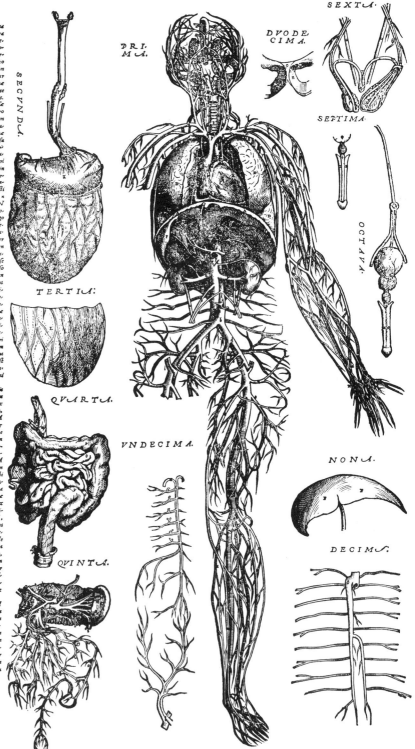

OMNES *figuræ hac charta impreſſæ ſimul ad unam ſpectant, quæ ex capite, aut uitæ commodius ducterix, illi figuræ agglutinanda uenit, quæ neruorū ſeriè proponit, ac in folio m inſcripto, ſeu omnium ultimo ſpectatur. Quod eos admonui uoluimus, qui m imparata incedine exemplaria, ſuoſp marte ac induſtria hæc ſibi concinnabunt: in quo opere cum in glutinando cp ſingulis à ſuperflua papyro reſcindendis, cum coloribus ſi uſum erit adhibendis, quiſp quantum uolet præſtabit. Deinut robori conſulatur, toti huic chartæ non inutiliter membrana ſubglutinabitur, priuſquam eo ea charta meret fruſta quot figuras complectitur, d'uidatur: quibus numerum in hoc aſcribam, ut quo quæq; loco committenda ſit, explicem, quantumq; in me eſt ſtudioſorum labori conſulam.*

PRIMAM *cæterorum præcipuam, ac uelut aliarum omniū baſim ex figura nudam muleris imaginem exprimentis proportione, uti cp cæteras omnes hic obiter delineauimus, quæ primam unicuiq; in ambitu proxime ad delineationem a reliqua charta eſt reſcanda, interim latiuſcula reſeruata in capitis uerticis portione, à qua poſtmodum g'urinari queat, ſimulaſque reliquæ partes ipſ fuerint committeſſe.*

SECVNDAE *figuræ ſtomachum cp anteriorē uentriculi ſedem ſimul cum ſuperiori membrana omenti, harumq; partium uaſis cp nerui proponent, aliquot aliæ. priuſquam prime committur, ueniunt agglutinandæ.* TERTIA *enim, quæ poſteriorem p'onem exprimit, tre ue inſerioris membranæ ſedis quæ ſub colo intestino conſiſtit, quà id uentriculo exporrigitur, ita ex proportione ſuperiori membranæ omenti, glutinanda eſt, ut omentum ſacculi imaginem referat.* QVARTA *dein, quæ inteſtinorum imaginem ſpectandam offert, tertiæ figuræ tergo ea ſede eſt committenda, qua inferius uentriculi orificium inteſtinorum principio continuatur. Sedem hanc f m ſecunda figura et quarta poſſū indicabt. Quum tamen quartam committis, conduxeris uering ad f laterum chartæ portionem aſſeruare, illamq; tertiæ figuræ, ut quarta uald.us hæreat, agglutinare. Nunc ſecunda prime nexuras, tranſuerſam ſectionē in tria imbiduces, ulti ſinhumil or ſep;ti tranſuersſ ſedi occurris, iecoris ſinū indicans, quo id ſto, morbo hoc tranſeantis ced t. Per hoc foramē ſtoma chum ita tranſmittes, ut æ ſinæ arteriæ ſubijcitur, uentriculuſq; ſua ſede conſtat. Nexum uerò ad prime figuræ tergum iuxta nu, per d'clum ſoramen mſderis.*

QVINTA *figuræ inferioris omenti membranæ port onem exprimit quæ poſt-riori uentriculi ſubijcitur ſedi, ac uenæ uortæ diſtributionem ſimul cum arterijs neruiſq; hac excurrentibus ſuſtinet. Inſuper præter eiuſmodi uaſa, hic et a ſien cum uenis ac arteriis per meſenteriū diffuſis ſpectatur, totaq; figura ad primam in pectoris cauo ita eſt gluti nenda, ut uiperit in utriſq; figuris occurrente a m urcem r. buciluer.*

SEXTA *figuræ uterum cum teſtibus ſe minal;bus cp naſci proportione, no ſiquam cæterorum medio circum ſciers, prime tubi in linis dextro latere et ſiniſtro iungetur, uti uena arterias ſeminalis ſimul eo um cp s m ſ niſtro latere in prime f gure committeſlatur. Quum hanc iungis, etiam* SEPTIMA, *quæ ueſicam ac umbilici uaſa ſimul cum meatuum urinam è renibus deſerentium portione, eademſedi ad primam ita eſt conglutinanda, ut meatus illi ſeminariis uaſis ſubglui ner um t, ipſa autem ueſica uterro incumbat, proporcione quā poſiu oti imam hic obſeruans, quam bari um delineatio æ continuitas facilè commonstrant.*

OCTAVA *iecoris gibbæ ſedis tam por;tiorum deli neat;m cort nens, quæ in anteriori cor t porissiqd oneſſ octani occurris, ti à iecoris fiſſum refert, cui uena ab umbilico ducta inſertur. Hanc ita z ucluti ex punctlo tantum, illic iecori in prima figura committes, ubi A inter F cp ſpectabis.*

NONA *uenæ pari carenis ſeriem oſtendens ad prime f gure trigum eſt committenda, ubi cæuæ cauæ, ſeu uenam illam ſine coniuge promit. atq; id et brompteſsiad o t mambabus figuris animū adh'bueris.*

DECIMA, *qua duas partes, ubi à ſuperflua papyro reſecta erit, conſti;uret, ſuperiori ſede uenam ac arteriam dextri lateris delineatas continnet, quæ ſub pectoris oſſe deorſum redentis, ſuperioremē chominis ſedem berlis, Hæc ſi ſuo charactere q m ingulo ad prime figuræ q glutinabitur. * aut;;mi ad remum prime figure, qui ad dextrum latus ż cp m reſectis conſpicuur. Hemilior decimæ figuræ pars uaſa oſtendt in inferiorum abdominis ſedi a excurrentia, quæ in ſii ad primam figuræ ſunt cūm ſtenda, ubi m dextro latere l iuxta uaſa crus prætenſa obuii eſt, illorumſq; uaſorū apparentu in via. Nunc primam figuram chartæ committes, quæ ner uorum ſeriem proponit, cp ſeu negoc o obſeruabis quo ſitu ſingula ſint locanda, ſi mod) characterum mdicem pariumą deſcriptionem operi adhibueris.*

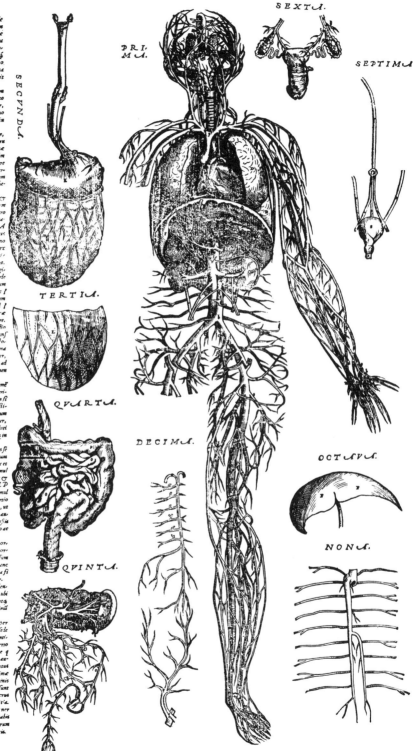

SECVNDA.

PRIMA.

SEXTA.

SEPTIMA.

TERTIA.

QVARTA.

QVINTA.

DECIMA.

OCTAVA.

NONA.